A NATURALIST'S GUIDE TO THE

REPTILES
& AMPHIBIANS
OF
BALI

A NATURALIST'S GUIDE TO THE
REPTILES
& AMPHIBIANS
OF
BALI

Ruchira Somaweera

JOHN BEAUFOY PUBLISHING

This edition published in the United Kingdom in 2020 by John Beaufoy Publishing Ltd
11 Blenheim Court, 316 Woodstock Road, Oxford OX2 7NS, England
www.johnbeaufoy.com

Photo captions and credits
Front cover: Tokay Gecko, Dog-toothed Cat Snake, Bowring's Supple Skink, Javanese Bullfrog
Back cover: Maned Forest Lizard
Title page: Boie's Kukri Snake
Contents page: Javanese Bullfrog

Photo credits: see page 176

The presentation of material in this publication and the geographical designations employed do not imply the
expression of any opinion whatsoever on the part of the Publisher concerning the legal status of any country,
territory or area, or concerning the delimitation of its frontiers or boundaries.

ISBN 978-1-912081-25-7

DEDICATION
To my wife Nilu and the two little monsters Rehan and Nehan for tolerating my long absences from home while
on extended holidays in Bali over several years. This book is a simple dedication to the three of you.

Edited by Krystyna Mayer
Designed by Gulmohur Press, New Delhi

Printed and bound in Malaysia by Times Offset (M) Sdn. Bhd.

·CONTENTS·

The Lesser Sundas White-lipped Pit Viper is among the most common snakes encountered in Bali.

Geography and Climate of Bali

Bali is a humid tropical island at the easternmost end of the Greater Sundas Islands group of Indonesia within the Indian Ocean. It is located between Java to the west (separated by the 3.2km-wide Bali Strait) and Lombok to the east (separated by the 20km-wide Lombok Strait). Bali is 153km long and 112km wide, with a land area of 5,577km².

Several volcanic mountain peaks more than 2,000m high lie from the centre to the eastern side across Bali. As a result, the island's surface geology is mostly composed of volcanic deposits and alluvial sediments, with a few limestone regions in the extreme west and south of the island. At 3,031m, above sea level (asl), the island's highest point is the still-active stratovolcano Mt (or Gunung) Agung. Permanent inland waters are scarce on the island. Lake Batur, a crater lake, is the island's largest still-water body, while the 75km-long Ayung River is its longest flowing water body. Bali is situated in the south-west corner of the Coral Triangle – the region of the highest marine biodiversity on the planet.

Bali's highest peak, Mt Agung.

Three small islands, Nusa Penida, Nusa Lembongan and Nusa Ceningan, lie 14km south-east of Bali separated by the Badung Strait. They are administratively a part of the Klungkung regency of Bali. The largest of the three, Nusa Penida, is 207km² in size, with the highest point of 529m at Gunung Mundi. The islands mostly consist of limestone (not volcanic rock as is the case for most of Bali) and comprise rolling hills. Two other much smaller satellite islands (Pulau Serangan and Pulau Menjangan) and several tiny islets are located much closer to Bali.

Limestone cliffs along the coast of Nusa Penida.

Bali and the Nusa islands are located entirely within the tropics, with the equator passing 8° N of the island. The climate is characterized by constant high temperatures (averaging 28° C in Denpasar) and high relative humidity (more than 85 per cent). Rainfall is present most of the year, but the wettest periods occur when the west monsoon visits the island's hills and the southern slopes in November–March. The central mountains receive an annual rainfall of 4,000mm. The Nusa islands and northern and western parts of Bali are drier than other areas.

Given the small size of the island, Bali's floral and vegetation diversity is relatively low. An estimated 3,000–3,500 species of vascular plant occur on the island, but endemism is very low. The vegetation of Bali can be broadly classified under three climatic systems:
1. **Tropical rainforest climate** with lowland and montane rainforests in the island's central part.
2. **Tropical monsoon climate** with moist and dry deciduous forests and monsoon forests in the lowlands.
3. **Savannah climate** with grassland, dry, open scrub and palm groves in the north and north-east.

Aquatic vegetation can be found within inland wetlands (streams, rivers and seasonal ponds) and coastal habitats (mangroves, salt marshes, seagrass beds, sand dunes and beaches, mudflats, lagoons and estuaries). Agricultural lands (monoculture stands, cropland, home gardens and paddy fields) also have their own vegetation communities and provide alternative habitats for herpetofauna. A few examples of some of the common habitat types are shown here.

Lowland rainforest at the foothills of Mt Batukaru.

Dry deciduous forests on limestone hills at Nusa Penida.

Grassland and palm groves at the western end of Bali.

Brackish water mangrove forest at Nusa Lembongan.

Freshwater lake at Bedugul.

Paddy fields in south-east Bali.

Permanent river flowing through secondary forests and mixed home gardens in Gianyar.

HERPETOFAUNA OF BALI

The Lombok Strait to the east of Bali marks the separation of two biogeographical regions, the Indomalayan subprovince to the west and the Australasian province to the east. This boundary line is termed the Wallace's Line, after the English naturalist Alfred Russel Wallace, who drew it in 1859. Thus Bali (and Lombok) is the easternmost limit for some of the species with an Asian origin, and the westernmost limit for a few with Lesser Sundanese/Australasian origins.

Nevertheless, Bali (and Java) among the main islands of the Sundas, supports the fewest number of species and has the lowest level of endemism among vertebrates. This could be due to the turbulent volcanic history of the islands, which may not have given enough time for endemism to develop.

Compared with neighbouring larger Indonesian islands, the diversity of amphibians of Bali is relatively poor, with 15 species known. This is partly due to the drier climate and limited availability and variety of inland waters. Only one species, the Bali Chorus Frog *Microhyla orientalis*, is endemic to the island. The Montane Chorus Frog *Oreophryne monticola* is the only member of this genus with a range within the Sunda Shelf.

Table Family-level Diversity of Amphibians in Bali

Family	No. of species
Megophryidae, Litter Frogs	1
Bufonidae, True Toads	2
Microhylidae, Narrow-mouthed Frogs	4
Dicroglossidae, Fork-tongued Frogs	4
Ranidae, True Frogs	3
Rhacophoridae, Asian Tree Frogs	1

By contrast, the diversity of reptiles is relatively high given the small size of the island, and several species new to science have recently been discovered and are awaiting description. As of July 2019, a total of 79 reptile species from 20 families are known to occur in Bali.

Table Family-level Diversity of Reptiles in Bali

Family	No. of species
Turtles	
Emydidae, New World Pond Turtles	1
Geoemydidae, Asian Hard-shelled Turtles	3
Trionychidae, Soft-shelled Turtles	1
Cheloniidae, Hard-shelled Sea Turtles	4
Dermochelyidae, Leatherback Sea Turtles	1
Lizards	
Agamidae, Agamid Lizards/Dragons	4
Gekkonidae, Cosmopolitan Geckos	7
Chamaeleonidae, Chameleons	unknown*
Lacertidae, Wall Lizards	1
Scincidae, Skinks	11
Dibamidae, Worm Lizards	1
Varanidae, Monitor Lizards	1
Snakes	
Acrochordidae, File Snakes	1
Pythonidae, Pythons	2
Xenopeltidae, Sunbeam Snake	1
Colubridae, Colubrid Snakes	24
Lamprophiidae, Sand Snakes, House Snakes and similar	2
Elapidae, Terrestrial Elapids**	5
Homalopsidae, Mangrove Snakes	2
Pareidae, Asian Slug Eaters	1
Viperidae, Vipers And Pit Vipers	1
Gerrhopilidae and Typhlopidae, Blind Snakes	3
Crocodilians	
Crocodylidae, True Crocodiles	1

* Believed to have been deliberately introduced but no information on the population status is available.
** No extended assessment of the sea snakes of Bali has been conducted, and the available records are limited to the two species of sea krait known from Bali's beaches. However, based on records from nearby islands, personal observations and expert opinions, it is considered that up to 18 species of true viviparous sea snake and an additional species of sea krait are likely to occur in the seas around Bali and Nusa islands.

CONSERVATION OF HERPETOFAUNA ON BALI

Reptiles play an important part in the beliefs of the Balinese people. For example, according to the ancient manuscript of Catur Yoga, the great serpent Antaboga created the giant turtle Bedawang Nala, then strapped the world onto its back with a pair of crested serpents, or Nagas. To this day, all earth tremors are believed to occur as the result of the shifting of the turtle and a disruption in the delicate balance in the harmony of the natural world.

However, despite the importance of herpetofauna in the Bali people's beliefs, most of the island's species are threatened by human activities. With more than 730 people per square kilometre Bali is one of the most densely populated islands in the world, resulting in significant pressures on its natural environment and wildlife.

The loss of suitable living habitat is by far the most serious threat to all herpetofauna on the island. This is due to deforestation, urbanization, agriculture, habitat alteration, forest fires, habitat fragmentation and large-scale development programmes. Destruction and development of marine habitats such as beaches, mangroves and coral reefs are also detrimental to marine reptile species, especially the sea turtles. Coral reefs and seagrass beds around Nusa Penida are being destroyed at an alarming rate to make space for seaweed farms, which are one of the main income generators for the locals.

Following the destruction of their natural habitats, most amphibians and reptiles have adapted their lifestyles to live in human-made habitats such as plantations, home gardens

Bedawang Nala and the Nagas at the base of a padmasana at Pura Puseh in Denpasar.

Many reptiles and amphibians in Bali have now adapted to live in disturbed human habitats. These include venomous species such as the Lesser Sunda White-lipped Pit Viper.

and even cities. This is clearly the case for most amphibians and lizards, especially geckos. However, within these new habitats the herpetofauna have become more prone to killing by humans, exposure to chemicals, mechanical damage by machinery, and predation by domestic cats, dogs, poultry and other animals such as crows. The fragmentation of natural habitats also has long-term negative impacts on genetic diversity and the continuation of populations in species that are not very mobile.

Compared with some other parts of Indonesia, consumption of amphibians, lizards and snakes by humans is not a significant threat to most herpetofaunal species in Bali. Few species are consumed for medicinal purposes. Oils extracted from cobras are considered to cure skin conditions such as eczema and burns, while meat and oil extracts from the Common Sun Skink are believed to cure allergies, ulcers and open wounds. Frogs of the family Dicroglossidae (especially *Fejervarya* species) are consumed locally, but no significant exports are recorded.

An exemption is the sea turtles. Capture and trade of sea turtles in Bali began in the 1960s and the industry peaked in 1978, making Bali one of the leading export centres of turtle goods alongside a few other Indonesian states, including Palembang, Jakarta, Surabaya, Pontianak and Ujung Pandang. More recently, sea turtles were declared protected and endangered, making it illegal to hunt them and to trade in them. Subsequent interventions by religious associations and conservation groups resulted in a much reduced number of turtles in trade. However, illegal harvesting of turtles and their

A few 'turtle centres' occur in Bali, some looking after injured animals and also conducting head-start programmes. Improperly managed and conducted head-start programmes make very little or no contribution to the conservation of turtles.

eggs still continues. Currently, turtles may be legally obtained for offerings, as in the *Manusa Yadnya*, the core human rituals at large-scale Balinese Hindu festivals including Padudusan Agung, Pancabali Krama, Ekadasa Rudra, Tri Buana and Eka Buana.

Although no formal assessments have been made yet, the introduced American Bullfrog *Aquarana catesbeiana* and Red-eared Slider *Trachemys scripta elegans* may have adverse impacts on native species through competitive exclusion and predation within the water bodies of Bali. Chytrid fungi and *Saprolegnia* have not been recorded to cause amphibian declines in Bali (or Indonesia), though both pathogens have been reported from the wild in the region.

Selling reptile body parts (such as ornaments made out of reptile skins) and live reptiles for the pet industry continues in Bali, especially at markets in Satria, Krenang and Beringkit. Tokay geckos, pythons and freshwater turtles are commonly sold alive, and turtle-shell products are not uncommon either.

Despite many species being threatened in the wild, only a handful of reptile species in Bali are legally protected. The list of protected flora and fauna in Indonesia in Regulation No. 7 of 1999 includes no amphibians and only seven species of reptile inhabiting Bali.

1. Loggerhead Turtle *Caretta caretta*
2. Green Turtle *Chelonia mydas*
3. Hawksbill Turtle *Eretmochelys imbricata*
4. Olive Ridley Turtle *Lepidochelys olivacea*
5. Leatherback Turtle *Dermochelys coriacea*
6. Saltwater Crocodile *Crocodylus porosus*
7. Burmese Python *Python bivittatus*

The IUCN Red List of Threatened Species (version 2016–2) lists 23 species from Bali: two as Endangered, three as Vulnerable and the rest as Least Concern. The definitions of the different IUCN categories are given in the Appendix (see p. 170).

1. Montane Chorus Frog *Oreophryne monticola*, Endangered*
2. Green Turtle *Chelonia mydas*, Endangered
3. Black Marsh Turtle *Siebenrockiella crassicollis*, Vulnerable
4. King Cobra *Ophiophagus hannah*, Vulnerable

5.   Burmese Python *Python bivittatus*, Vulnerable
6.   Crested Toad *Ingerophrynus biporcatus*, Least Concern
7.   Cricket Frog *Amnirana nicobariensis*, Least Concern
8.   White-lipped Frog *Chalcorana chalconota*, Least Concern
9.   Blue-tailed Snake-eyed Skink *Cryptoblepharus renschi*, Least Concern
10. Asian Grass Lizard *Takydromus sexlineatus*, Least Concern
11. Great Crested Canopy Lizard *Bronchocela jubata*, Least Concern
12. Asian Water Monitor *Varanus salvator*, Least Concern
13. Dog-toothed Cat Snake *Boiga cynodon*, Least Concern
14. Malayan Krait *Bungarus candidus*, Least Concern
15. Cuvier's Reed Snake *Calamaria schlegeli*, Least Concern
16. Short-tailed Reed Snake *Calamaria virgulata*, Least Concern
17. Paradise Tree Snake *Chrysopelea paradisi*, Least Concern
18. Southern Indonesian Spitting Cobra *Naja sputatrix*, Least Concern
19. Common Southeast Asian Tree Frog *Polypedates leucomystax*, Least Concern
20. Spotted Keelback *Rhabdophis chrysargos*, Least Concern
21. Striped Litter Snake *Sibynophis geminatus*, Least Concern
22. Lesser Sunda White-lipped Viper *Trimeresurus insularis*, Least Concern
23. Javan Keelback *Xenochrophis melanozostus*, Least Concern

*Suggested to be downgraded to Least Concern as recent surveys have found several new populations expanding its known distribution.

Bali is a signatory to the CITES (Convention on the International Trade of Endangered Species), and commercial trade of species between Indonesia (including Bali) and other countries has been restricted. The following species from Bali are listed in CITES. The definitions of the different CITES Appendices are given in the Appendix (see p. 170).

1.   Saltwater Crocodile *Crocodylus porosus*, Appendices I and II
2.   Loggerhead Turtle *Caretta caretta*, Appendix I
3.   Green Turtle *Chelonia mydas*, Appendix I
4.   Southeast Asian Soft Terrapin *Amyda cartilaginea*, Appendix II
5.   South Asian Box Turtle *Cuora amboinensis*, Appendix II
6.   Black Marsh Turtle *Siebenrockiella crassicollis*, Appendix II

WHAT THIS BOOK IS AND WHAT IT IS NOT

This work is a general introduction and non-technical guide to the herpetofauna inhabiting Bali and the Nusa islands. It covers all the known herpetofauna to date (July 2019) from these islands, and also shows a few species that are likely to occur but have not yet been recorded. It is designed as a compact field handbook for non-technical users, and features brief descriptions that may assist in identifying the species in the wild. The book should be used in conjunction with existing, more-detailed technical books and scientific papers.

▪ WHAT THIS BOOK IS AND WHAT IT IS NOT ▪

FAMILY ACCOUNTS

The family accounts describe in brief the global number of species included in each family, and the broad and common morphological characteristics of its members; they also provide a summary of the natural history of each group, and the number of species occurring or probably occurring on Bali and Nusa islands. Families are listed in phylogenetic order following The EMBL/EBI Reptile Database (www.reptile-database.org).

SPECIES ACCOUNTS

Species are listed in alphabetic order of their scientific names within a family. Each account includes:

- Scientific name. As a universal rule, each species has a unique, two-worded name (referred to as the 'binomial name') written in Latin and italicized. The first part of the name identifies the genus to which the species belongs, the second part identifies the species within the genus. A given species only has one current scientific name.
- Common names in Indonesian, Balinese and English (where available). A given species can have multiple common or 'vernacular' names depending on the geographic location, colour phase and sometimes even the life stage. Additionally, communities in different parts of Bali (and Indonesia, and the world) may refer to the same species by different names based on the language, dialect, ethnic and cultural backgrounds.
- Average length: Shell/Carapace Length; SVL – Snout to Vent Length); or TL – Total Length, depending on which group an animal belongs to. Where available, the average lengths of females (♀) and males (♂) are given separately.
- Diagnostic features (including for tadpoles).
- Habitats in which a species is usually encountered.
- Natural history of the species.
- In the case of some reptiles, the level of danger a particular species poses to humans.

  Because information on the distribution of different species of herpetofauna in Bali is scarce and largely incomplete, it is not included.
  Care should be taken when interpreting the danger levels of each species. Several venomous species may not pose a threat to humans as their behaviours and/or morphology may not allow an effective bite. On the other hand, some non-venomous species may cause damage through their bites (from the teeth).

COLOUR ILLUSTRATIONS

All species are illustrated with colour photographs of live animals, unless a species is extremely rare or known from only a handful of specimens or even from just one specimen; in such cases, pictures of the preserved specimens are included. Where possible within the very limited space, several colour variants, depending on age and sex, are shown. Photographers are credited on p.176

MORPHOLOGY & SCALATION

FROGS AND TOADS

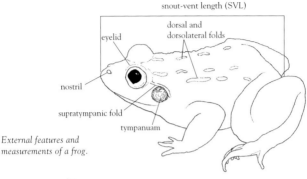

snout-vent length (SVL)

dorsal and
dorsolateral folds

eyelid

nostril

supratympanic fold

tympanuam

*External features and
measurements of a frog.*

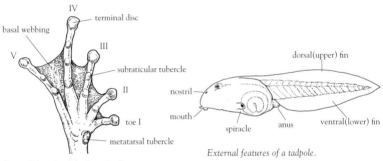

IV

terminal disc

basal webbing

III

V

subraticular tubercle

II

toe I

metatarsal tubercle

Parts of the sole of the foot of a frog.

dorsal(upper) fin

nostril

mouth

spiracle

anus

ventral(lower) fin

External features of a tadpole.

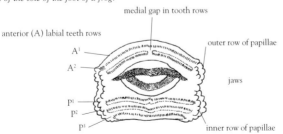

medial gap in tooth rows

anterior (A) labial teeth rows

outer row of papillae

A^1

A^2

jaws

P^1
P^2
P^3

inner row of papillae

posterior (P) labial teeth rows

Mouthparts of a tadpole.

TURTLES

nuchal/cervical

gular

humeral

inframarginal

pectoral

costals

vertebrals

shell length

abdominal

marginals

femoral

marginal

anal

Upper shell scutes

Lower shell scutes

LIZARDS

snout-vent length (SVL)

total length TL

Measurements of a lizard

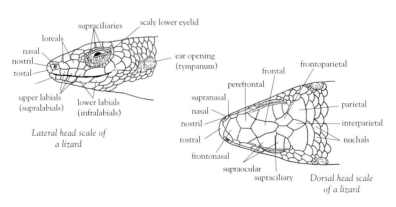

supraciliaries

scaly lower eyelid

loreals

nasal

nostril

rostal

ear opening
(tympanum)

upper labials
(supralabials)

lower labials
(infralabials)

*Lateral head scale of
a lizard*

frontal

frontoparietal

perefrontal

supranasal

parietal

nasal

interparietal

nostril

rostral

nuchals

frontonasal

supraocular

supraciliary

*Dorsal head scale
of a lizard*

SNAKES

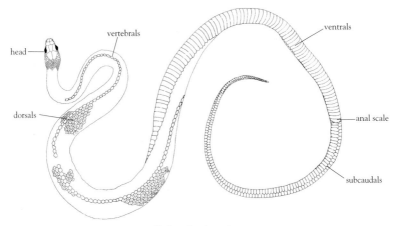

Body scales of a snake

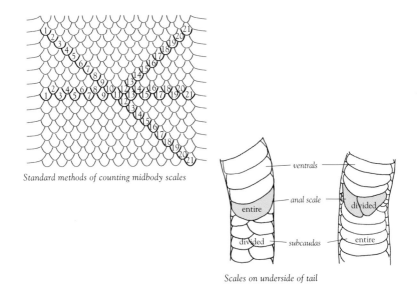

Standard methods of counting midbody scales

Scales on underside of tail

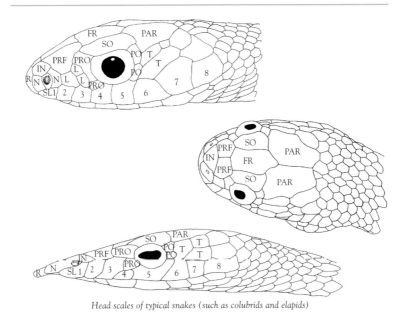

Head scales of typical snakes (such as colubrids and elapids)

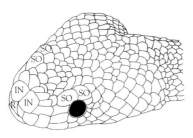

Head scales of a viper

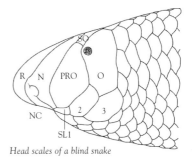

Head scales of a blind snake

KEY TO SNAKE HEAD SCALES

FR frontal, IN internasal, L loreal, N nasal, NC nasal cleft, O occular, PO postoccular, PRF prefrontal, PRO preoccular, R rostral, SL suparlabials, SO supraoccular, T temporals

▪ First Aid ▪

First Aid for Snake Bite

Antivenom is the only specific treatment available for venomous snake bites, so as soon as proper first aid is applied, the victim should immediately be taken to a hospital with antivenom and other facilities to treat complications. Here are some dos and don'ts for treating a snake-bite victim.

DOS FOR A VENOMOUS SNAKE BITE

- Keep the affected person calm, and assure them that bites can be effectively treated. Venom spreads faster if the heart beats faster and a victim is moved, than if they are kept still.
- Apply a broad elastic pressure bandage, first wrapping over the bite wound, then the whole limb starting below the bite site, wrapping it gradually (for example from fingertips to armpits, even if the bite is on the palm, or from the toe-tips to the groin even if the bite is on the knee). The pressure should be kept moderate, as if wrapping a sprained ankle, and should not cut off the blood flow. Attach a rigid object such as a stick or folded book as a splint, and further bandage it to the wrapped limb to immobilize it. Note, however, that this pressure-immobilization method may not be very effective against bites from vipers.
- If the bite is on the trunk, neck, face or head, apply and maintain firm pressure on the bite site (for example by keeping the bite site pressed by the hand). It is difficult to apply the pressure-immobilization method to these parts effectively.
- Keep the affected area below the heart level to reduce the flow of venom towards the heart, then to other body parts.
- If possible, monitor the person's important signs – temperature, pulse, rate of breathing and so on. If there are signs of shock, lay the person flat, raise the feet about a foot, and cover the person with a blanket.
- Give only paracetamol, if there is pain or fever.

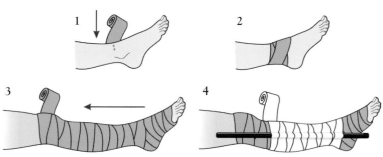

Correct application of the pressure-immobilization method.

- Remove rings, bracelets, watches and so on, as if swelling takes place it will be painful to remove these.
- Transport the person to the nearest hospital as soon as possible, keeping them on a stretcher or chair.

DON'TS IN A VENOMOUS SNAKE BITE

- DO NOT wait to see if the bite causes any problems – treat it straight away.
- DO NOT allow the person to panic and run around – doing so increases the heart rate and in turn increases the spread of the venom. Always carry the patient.
- DO NOT attempt to cut out or suck out the venom from the bite wound.
- DO NOT wash the wound as a venom swab may be needed for correct identification of the snake.
- DO NOT give aspirin (paracetamol is fine if there is pain or a fever), fruit juices (anything with high levels of potassium like king coconut) or alcohol. Some of these increase the absorption of venom and may also harm the heart and increase kidney damage.
- DO NOT put tourniquets or ties above the bite wound. These would cut the blood supply for the whole limb, causing severe tissue damage.
- DO NOT apply cold compresses or ice to a snake bite.
- DO NOT apply local remedies. Most of these would cause more harm than good.
- DO NOT waste time (or your life) searching for or trying to kill the snake. Modern medicine has other means to identify the snakes responsible for a bite and patients are treated according to their symptoms, *not* the species of snake that bit them.

DOS AND DON'TS IF SPITTING COBRA VENOM GETS IN THE EYES

- Wash the eyes with plenty of water immediately by either dipping the face in a large bucket of water, or holding it under a running tap. Keep the eyes open when washing and blink. Continue for 15 minutes plus.
- DO NOT put any traditional remedies (such as tamarind leaf juice) on the eye. However, other benign fluids such as milk can be used.
- DO NOT rub or blot the eye to try to remove the venom.

GLOSSARY

adult Sexually mature individual.
aerobic Relating to, involving or requiring free oxygen.
albino Congenital lack of normal colour pigmentation in skin.
ambient temperature Overall temperature of environment surrounding an animal.
anterior Pertains to near the front/towards the head.
anus Opening at end of digestive tract through which solid waste or excrement (stools) leaves the body. In lizards, located at base of tail.

apical pits Minute epidermal depressions found on or near the apex of dorsal scales of some snake species.

aquatic Lives in water.

arboreal Lives on trees or other vegetation.

autotomy Voluntary shedding of body parts of animals, usually in defence. Autotomy of the tail is common in many lizards.

basal At or near base.

canopy Upper layer or habitat zone formed by mature tree crowns and including other biological organisms such as epiphytes and vines.

canthus rostralis More or less angular ridge from anterior border of eye to nostril in reptiles and amphibians.

carapace Upper shell of a turtle.

carnivore Animal that hunts and eats other animals; meat eater.

caudal Of the tail.

chevron Inverted, 'V'-shaped mark.

cloaca Common chamber into which urinary, digestive and reproductive canals discharge their contents, and which opens to exterior through anus.

clutch Total number of eggs laid by a female at one time.

courtship Behaviour preceding mating.

cranial Often refers to top of head (cranial crest).

crepuscular Becomes active at dusk, dawn and twilight.

crest Longitudinal row of elevated scales along neck, back and/or tail.

cryptic Secretive or concealed by means of behaviour, colouration and/or patterning.

cutaneous Relating to or affecting the skin.

depressed Flattened from top to bottom (dorso-ventrally flattened).

dermal Of the skin.

dewlap Loose flap of skin under throat.

dimorphism Occurrence of two forms from various differences like colour and size. Often refers to sex-related differences.

diurnal Active during the day.

dorsal Of or pertaining to back or upper surface.

dorsolateral Pertains to upper sides.

dorsum Entire upper surface of an animal.

durophagy Eating behaviour of animals that consume hard-shelled or exoskeleton-bearing organisms (such as skinks among reptiles).

ecology Scientific study of interactions of organisms and their environment.

endemic Restricted/confined to specific region or area.

epidermis Protective outer layer of the skin.

femoral pores Small opening on underside of thigh of some lizards.

fossorial Living underground (in burrows or digs).

genus (pl. **genera**) Taxonomic category ranking below family and above species, and generally consisting of group of species exhibiting similar characteristics. In taxonomic nomenclature the genus name is used either alone or followed by a Latin adjective or

▪ GLOSSARY ▪

epithet, to form the name of a species.
granules Very small, flat scales.
gravid Female bearing eggs or embryos.
groin Area between abdomen and upper thigh on either side of body.
gular On or pertaining to throat.
habitat Environment an organism lives in.
hemipenis Copulatory organ of male reptiles.
herbivore Plant eater, vegetarian.
herpetile Reptiles and amphibians together.
herpetology Scientific study of reptiles and amphibians.
hybrid Crossbred animal (or plant). Offspring of two different species.
imbricate Overlap.
juvenile Not yet sexually mature.
juxtaposed Contiguous or non-overlapping.
keel Narrow longitudinal ridge on a scale.
keratinize Change to form containing the horny tissue keratin.
labial Of or pertaining to the lip.
lamella (pl **lamellae**) Pads under digits in some lizards.
lanceolate Shaped like a lance head (narrow oval shape tapering to point at each end).
lateral Of or pertaining to side.
loreal Space between nostril and eye.
mandible Jaw or jawbone.
mental Of the chin.
mental groove Groove located in middle of lower jaw on snakes, comprising loose skin that allows jaws to expand to accommodate prey larger than gape of snake's mouth.
mesoplastron Dermal bone present in plastron of certain turtles.
midbody scales Scales counted around middle of body (see illustration, p. 20).
middorsal Of or pertaining to middle of back.
midventral Of or pertaining to middle of belly.
nektonic Pertaining to organisms that actively swim in water column.
nocturnal Active at night.
nuchal Of back of head.
ocelli Eye-like and ring-shaped spots.
omnivore Animal that eats both meat and vegetables.
osteoderms Bony deposits forming scales, plates or other structures in dermal layers of skin.
oviparous Animal that lays eggs that later hatch.
oviposition Laying of eggs.
ovoviviparous Animal that holds eggs inside body until they hatch and living young are delivered. However, the embryo is sustained by the contents of the egg (yolk), not by any connection to the animal that holds it.
parthenogenesis Development of an individual from an egg without fertilization. This system allows the production of offspring by a female with no genetic contribution from a male.

patagium Membranous, wing-like structure.

pelagic Inhabiting open ocean.

plastron Bottom shell of turtle

posterior Pertaining to rear.

pores Small, hole-like openings on scales of some lizards and other reptiles.

preanal pores Pores situated in front of cloaca in lizards.

prehensile tail A tail capable of wrapping or grasping.

proboscis Long and mobile nose such as trunk of elephant.

reticulated Net-like pattern.

rostral Of the snout.

scansor Pads under digits in geckos.

scute Horny, chitinous or bony external plate or scale, as on shell of turtle or underside of snake. Also called **scutum**.

semiaquatic Able to live on both land and water.

setae Stiff, hair-like or bristle-like structure.

species Group of living organisms comprising similar individuals capable of interbreeding and producing fertile young. The 'species' is the principal natural taxonomic unit, ranking below a genus and represented by a Latin binomial name.

subadult Developmental life stage when an animal exhibits most but not all traits of a sexually mature adult.

subcaudal Scales below tail.

subspecies Population of a species occupying a particular geographic area, or less commonly, a distinct habitat, capable of interbreeding with other populations of the same species. Subspecies typically display intergrading physical differences.

supratympanal/supratympanic Positioned above tympanum (external eardrum).

suture Groove between non-overlapping scales.

taxon (plural **taxa**) a group of one or more populations of an organism or organisms seen by taxonomists to form a unit.

terrestrial Lives on the ground (land).

thermoregulate To regulate body temperature. Most reptiles cannot produce their own body heat and must rely on external or environmental heat sources. They control their core body temperatures by moving in and out of areas with varying temperatures and humidity levels.

truncate Shorten (as by cutting off top or end).

tubercule Knob-like projection.

tympanum External eardrum.

vent Anus, the opening from the cloaca to the outside.

venter Entire undersurface of an animal.

ventral Of or pertaining to underside, the lower surface.

ventrals Belly scales.

vertebrals Of the spine.

vomerine teeth Specialized structures in a frog's mouth that help it to grip prey. They are not involved in chewing or killing prey, but only in restraining it.

MEGOPHRYIDAE
LITTER FROGS, HORNED FROGS AND RELATIVES

This group is represented by about 225 species naturally occurring in South and Southeast Asia.

The group is morphologically highly diverse, so it is difficult to make generalizations about the species. They usually have notably large, wide mouths and paddle-shaped tongues. Many members have spade-like projections on the feet. Some mimic dead leaves with projections on their snouts and over the eyes, so are well camouflaged against the forest floor background.

Litter frogs walk rather than jump when moving. They breed in pools within streams and have highly aquatic tadpoles. Tadpoles in stagnant waters have funnel-shaped mouths to feed at the surface, while those in fast-flowing waters have sucker-like mouthparts to cling on to rocks.

A single species of megophryid is known from Bali.

Megophryids such as this Hasselt's Litter Frog are well adapted to a terrestrial life.

Hasselt's Litter Frog ■ *Leptobrachium hasseltii* SVL 5cm
(*Bahasa Indonesia* Katak Hasselt, Katak Serasah)

DESCRIPTION *Physique* Body medium in size; head wider than body or equal in width; very large eyes; skin fold above tympanum; toes webbed at bases and with rounded tips. *Colouration* Dorsum shades of purple-brown or dark brown, with black or darker mottling; sides of face darker; limbs have dark and usually narrow cross-bands; eyes dark brown or black; venter light brown or cream coloured, with white speckles; juveniles bluish in colour. **TADPOLE** Free swimming; globular body; spiracle opening on side of body; 1 row of papillae along upper lip; more than 4 rows of labial teeth each on upper and lower lips; tail 1.5 times body length, shaped like a narrow leaf with a blunt tip; body brownish or greyish, sometimes without darker mottling with age; lighter markings on face and sides. **HABITAT AND HABITS** Most common in midland forested areas. Strictly nocturnal. Terrestrial and spends most time in moist leaf litter on forest floor. Advertisement call is a repeated *kir kir kir kir* sound that sometimes progressively increases in pace and pitch.

Adult colouration.

BUFONIDAE
TRUE TOADS

This large group is represented by about 620 species. It occurs on all continents except Antarctica. In Australia the family is represented by the introduced and invasive Cane Toad *Rhinella marina*.

True toads have thick, dry, warty skin, and some species have characteristic parotoid glands behind the eyes. These glands produce milky, latex-like alkaloid poisons (commonly termed 'bufotoxins'), which have a defensive function (it is highly advisable to wash your hands well if in contact with these secretions). Bufonids lack teeth in both the upper and lower jaws. The limbs are shortened, and adapted to walking or hopping. The toes do not end in discs. Internally, males have a Bidder's organ, which under the correct conditions transforms into an active ovary, causing a male toad to become a female.

Bufonids are nocturnal and predominantly terrestrial. At least one species, the Aquatic Swamp Toad *Pseudobufo subasper*, found in the Malay Peninsula, Borneo and Sumatra, is aquatic.

Reproduction is very variable in bufonids, with species including toads that lay eggs in single or paired strings, which hatch into tadpoles in water, terrestrial direct developers, and toads that give birth to live toadlets (true viviparity).

Two bufonid species are known from Bali.

Warty and dry skin, bony ridges on the head and parotid glands behind the eyes are characteristics of bufonids such as this Asian Spined Toad.

Asian Spined Toad
■ *Duttaphrynus melanostictus* SVL ♀ 10cm, ♂ 6cm
(*Bahasa Indonesia* Kodok Buduk, Kodok Puru, Bangkong Kolong)

DESCRIPTION *Physique* Body medium in size and stocky, with males in some populations much smaller than females; head with elevated bony ridges on snout, long dark crest bordering eyelids and another from eye to parotoid gland; parotoid gland large and elliptical/oval, 2½–3 times size of eye; tympanum very distinct; dorsum with two rows of 8–9 enlarged tubercules, usually tipped with dark brown spines; toes half to two-thirds webbed, except for 4th toe; males develop corn-shaped nuptial pads on fingers during mating season; juveniles lack warts and often have a very inconspicuous tympanum. *Colouration* Highly variable dorsum, comprising uniform shades of brown, orange, grey or even black; darker spots and streaks present in some individuals; venter uniform dirty white, with light brown speckles, mostly on throat and chin; throat of male turns to orange/yellow during mating season; juveniles more reddish than adults. **TADPOLE** Free swimming; globular, small body; spiracle opening on side of body; papillae only on corner of mouth; lowest row of labial teeth on bottom jaw different in length to other rows of labial teeth; tail has broad dorsal fin and narrow ventral fin with rounded tip; uniform dark in colour. **HABITAT AND HABITS** Recent introduction/immigrant to Bali. Even in the

1950s it was only found in Negara, West Bali. The species is now extremely abundant and widespread in Bali, especially in human habitats. Nocturnal, spending the day hiding under logs and stones, and in crevices, sometimes in groups. Hunts on the ground and preys on wide variety of invertebrates. Can occur in high concentrations, especially around artificial light in human habitats. Breeding often peaks at start of wet season and takes place in home-garden ponds, shallow pools and even paddy fields. Call comprises a repetitive *curr curr curr*, sometimes ending in a high-pitched whistling note. Eggs laid in double jelly strings.

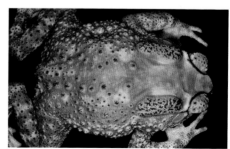

Adult colourations, and shape of parotid gland and ridges on head (bottom right).

Crested Toad ■ *Ingerophrynus biporcatus* SVL ♀ 9cm, ♂ 5.5cm
(*Bahasa Indonesia* Enggung, Kodok Bersurai Kecil, Kodok Puru Hutan; *English* Sunda Ridge-headed Toad)

DESCRIPTION *Physique* Body medium in size but stout; males much smaller than females; head has two elevated and elongated supraparietal ridges between eyes; parotoid gland small but distinct, more or less oval in shape; tympanum very distinct, circular or somewhat elliptical; toes half to two-thirds webbed, except for 4th toe; dorsum and flanks with dense tubercules; males develop corn-shaped nuptial pads on fingers during mating season. *Colouration* Variable dorsum comprising uniform shades of brown, grey or reddish, mottled with darker blotches and spots; venter uniform dirty white or with dark brown speckles, mostly on chest; throat of male turns to orange/red during mating season. **TADPOLE** Free swimming; globular, small body; spiracle opening on side of body; papillae only on corner of mouth; lowest row of labial teeth on bottom jaw is same in length as other rows of labial teeth; tail with broad dorsal fin and

narrow ventral fin with rounded tip; blackish on dorsum and pale venter; dusky tail fins. **HABITAT AND HABITS** Found up to 1,400m asl in both natural and human habitats. Nocturnal, and spends the day hiding under logs and stones, and in crevices. Hunts on the ground and preys on wide variety of invertebrates. Breeding takes place in shallow, stagnant water. Advertisement calls usually consist of 4–10 very short, complex and fast-phased *kur kur kur* pulses. Eggs laid in double jelly strings.

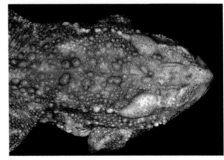

Adult colourations, and shape of parotid gland and ridges on head (bottom right).

MICROHYLIDAE
Narrow-mouthed Frogs

This large group is represented by close to 680 species. Microhylids naturally occur in South and Southeast Asia, the Americas, northern Australia, sub-Saharan Africa and Madagascar.

Most Asian members generally have small, teardrop-shaped, globose bodies with substantially smaller heads, short snouts and small mouths. The front limbs are short, and most have long back limbs that are adapted for digging and walking. Very few members are arboreal and have expanded toe discs. Most can inflate their bodies when sensing danger, and also secrete thick, sticky mucus from skin glands on the body.

This group shows a diverse array of reproductive strategies. Most species produce a large number of small eggs that float on water and hatch into tadpoles that are filter feeders in the water column. Some species have non-feeding tadpoles. Several lay eggs in water-filled tree-holes, and at least one breeds in water collected within bromeliads on trees. Other species lay eggs on moist forest floor and show direct development. Males of some species guard the eggs.

Four species are known from Bali.

A globose body, substantially smaller head and short snout are characteristics of most microhylids such as this Javanese Bullfrog.

Javanese Bullfrog ■ *Kaloula* cf. *baleata* SVL ♀ 7cm, ♂ 5cm
(*Bahasa Indonesia* Kintel Lekat, Belentuk; *English* Brown Bullfrog, Flowerpot Toad)

DESCRIPTION *Physique* Body small, round; limbs short; toes webbed and with spoon-shaped, blunt fingertips; toes slightly webbed; pair of prominent ridges on sole of foot. *Colouration* Dorsum brownish with darker and lighter mottling, and (occasionally) white-tipped tubercules giving a granular appearance; some individuals have a lighter band from behind eye over shoulder to elbow; inguinal spot at groin brick-red or yellow depending on geographic location; venter whitish. **TADPOLE** Free swimming; spiracle under middle of body and opening in a long tube; tail long, twice as long as body, and ending in fine filament; dorsum dark brown, greyish-brown or black with transparent tails. The tadpoles are specialized suspension feeders and thus lack keratinized mouthparts. **HABITAT AND HABITS** Found in primary and secondary forests as well as on agricultural lands. Mostly fossorial and secretive, emerging from refuge spots after strong and continued rainfall. However, also a good climber and is occasionally found in tree-holes, even on those more than 6m above the ground. Breeds during months of heavy rainfall, and tadpoles take about 2 weeks to metamorphose. Males often seen calling while floating at the water's surface. Call variable, with the common one being a repeated and short *bong* sound, like that from a Balinese *gamelan*.

Adult colouration.

Bali Chorus Frog ■ *Microhyla orientalis* SVL ♀ 3cm, ♂ 2cm
(*Bahasa Indonesia* Katak Bali Bermulut Kecil)

DESCRIPTION *Physique* Body very small, plumped; head small and narrow, and lacking supraciliary tubercule on upper eyelid; tympanum hidden by skin; limbs short; toes webbed at bases only, and with spoon-shaped fingertips; dorsal median grooves on toe discs; inner and outer metatarsal tubercules small. *Colouration* Dorsum light brown or pinkish-brown; sides of body darker; symmetrical, arrow-shaped, broad and darker mark on back; thin, discontinuous, pale line along vertebral line; black lateral stripe from above arm to half length of trunk; limbs barred with dark bands. **TADPOLE** Free swimming; spiracle under middle of body; lower lip expanded into a funnel shape, enabling it to feed at the water's surface; body and head flattened above; tail long, twice as long as body, lanceolate and ending in fine central filament; dorsum light brown with darker mid-dorsal band; venter grey; middle of tail with black marks dotted with gold. **HABITAT AND HABITS** Found in more or less wet and moist forests and agricultural lands in midlands and uplands. Fossorial, and presumably feeds on ants and termites. Advertisement call consists of series of high-pitched *krrk* notes, each emitted at a short interval. Males often seen calling from edge of water.

Adult colouration.

Palmated Chorus Frog ■ *Microhyla palmipes* SVL ♀ 2.5cm, ♂ 1.5cm
(*Bahasa Indonesia* Percil Berselaput, Percil Kaki Selaput, Katak Bermulut Kecil)

DESCRIPTION *Physique* Body very small, plumped; head small and narrow, with supraciliary tubercule on upper eyelid; tympanum hidden by skin; limbs short (but rear limbs comparatively longer than in Bali Chorus Frog, see opposite); toes webbed up to two-thirds or three-quarters of length, and with spoon-shaped fingertips; no dorsal median grooves on toe discs; inner and outer metatarsal tubercules small. *Colouration* Dorsum shades of brown or chestnut; lighter on sides; sides with black flecks; darker or black

spots sometimes make a collar and cross-bands on limbs; venter whitish. **TADPOLE** Free swimming; spiracle under middle of body, but covered by flap of skin; body and head flattened above; lower lip not expanded into funnel shape; tail long, twice as long as body, lanceolate and not ending in fine central filament; dorsum dark brown; tail pale. **HABITAT AND HABITS** Found in wet and moist forests and on agricultural lands in midlands and lowlands. Fossorial, and presumably feeds on ants and termites. Advertisement call a fast, high-pitched, repeated *kriiik*. Males often seen calling from vegetation at edge of water.

Adult colourations.

Montane Chorus Frog ■ *Oreophryne monticola* SVL ♀ 2.5cm, ♂ 1.5cm
(*Bahasa Indonesia* Percil Gunung)

DESCRIPTION *Physique* Body very small, plumped; head small; fingers and toes with circular discs at tips and unwebbed; scattered tubercules on back; tympanum hidden under skin.

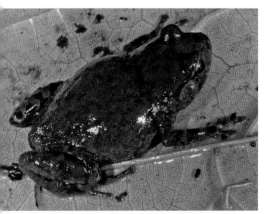

Colouration Highly variable. Dorsum shades of brown, black, purplish, yellow, orange or red; some individuals have darker or lighter mottled marks; venter whitish or cream coloured. **TADPOLE** No free-living tadpole stage. Tadpoles develop directly into froglets inside eggs before hatching. **HABITAT AND HABITS** Inhabits rainforest and montane habitats in highlands and midlands, as well as other well-vegetated habitats (like Eka Karya Botanical Gardens). Males call from forest floor. Presumably lays eggs among moist litter or amid aerial tubers of epiphytes.

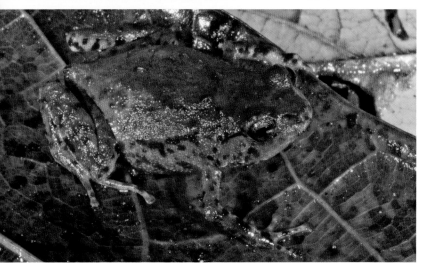

Adult colourations.

▪ FORK-TONGUED FROGS ▪

DICROGLOSSIDAE
FORK-TONGUED FROGS

This group is represented by about 215 species that naturally inhabit tropical and subtropical regions of Asia and Africa.

It is a morphologically diverse group, but most members have a generalized frog body plan with a moist skin that lacks hard tubercules or ridges. They range vastly in size. Males of some genera have fang-like structures on their lower jaws, which are presumably used in territorial combats. Species that burrow have spade-like tubercules on their back limbs.

Most members are terrestrial or semiaquatic, with a strong liking for places close to water, while some species, like the Common Puddle Frog *Occidozyga laevis*, are highly aquatic. A few species burrow into loose soil. Larger species take vertebrate prey including other frogs.

Reproduction within this family is extremely variable. Some species make scrapes at the sides of streams, with the males guarding the eggs and after hatching even transporting the tadpoles to water on their backs. No direct development is known in this group. The free-living tadpoles of most species are highly carnivorous.

Four species are known from the islands, while additional species such as the Javan Giant Frog *Limnonectes macrodon* may occur there (although there are no confirmed records of this species from these islands).

Large-bodied dicroglossids such as these East Asian Bullfrogs Hoplobatrachus rugulosus are commonly collected for human consumption in Southeast Asia. This species is not known from Bali.

Crab-eating Frog ■ *Fejervarya cancrivora* SVL ♀ 12cm, ♂ 9cm
(*Bahasa Indonesia* Katak Sawah, Katak Hijau, Katak Hutan Bakau; *English* Mangrove Frog)

DESCRIPTION *Physique* Body medium in size but robust; head narrow with pointed, oval snout, a rounded canthus rostralis and small glandular warts on sides; nostrils oval, located closer to tip of snout than eye, and with small flap; dorsum smooth and with interrupted but short dorsolateral folds; flanks with glandular warts and folds; tympanum distinct and with supratympanic fold; toes more than half webbed and with dermal fringes; 1 metatarsal tubercle; males develop nuptial spines on nuptial pads. *Colouration* Dorsum greyish-brown, greenish-brown or cream, with irregular darker markings on head (inter-orbital bars), dorsum and flanks, and darker bars on limbs and lips; venter whitish, occasionally with scattered darker mottlings, especially on throat. **TADPOLE** Free swimming; oval body; spiracle opening on side of body; 1 row of papillae along upper lip; less than 4 rows of labial teeth each on upper and lower lips; tail less than twice body length with a much broader upper fin than lower one; dorsum dark grey with darker spots. **HABITAT AND HABITS** Inhabits forests and agricultural lands in midlands and lowlands, as well as mangroves and other brackish water habitats. Tolerates wide range of salinities and is the only species of living amphibian that can inhabit saline water constantly. Diet ranges from predominantly crustaceans (in brackish water populations), to arthropods and smaller vertebrates (including snakes) in more inland populations. Reproduction takes place year round, but mostly at beginning of wet season. Advertisement call sounds like fast, deep throat gargle.

Adult colourations.

Paddy Field Frog ■ *Fejervarya limnocharis* SVL ♀ 5.5cm, ♂ 4.5cm
(*Bahasa Indonesia* Katak Sawah, Katak Tegalan, Godogon)

DESCRIPTION *Physique* Body small; head narrow with pointed snout projecting beyond mouth; dorsum with small tubercules or sometimes small, longitudinal folds; thighs granular; tympanum distinct and half to two-thirds the diameter of eye; toes with blunt and slightly swollen tips, half webbed, subarticular tubercules small and prominent; 2 metatarsal tubercles; males with pad-like subdigital tubercules under first finger. *Colouration* Dorsum

greyish-brown, greenish-brown or olive marbled with darker markings; thin or thick yellowish/orangey vertebral line from snout to groin present in most individuals; limbs and lips barred; thighs laterally yellowish, and some individuals have light line along calf; usually 'V'-shaped dark mark between eyes; venter white; throat of male mottled with brown and sometimes with dark 'M'-shaped band. **TADPOLE** Free swimming; body oval and long with front half flexed forwards and upwards; spiracle opening on side of body; 1 row of papillae along upper lip; less than 4 rows of labial teeth each on upper and lower lips; tail about twice length of body, gradually tapering, acutely pointed; dorsum dark olive or grey with darker speckling; end of tail sometimes with darker spots or bars. **HABITAT AND HABITS** Semiaquatic and commonly encountered close to freshwater bodies, including within urban areas, plantations and paddy fields. Breeds during the rains, and eggs are attached to vegetation in batches. Advertisement calls highly variable, but usually consist of rapidly repeated *eeep*.

Adult colourations.

Green Puddle Frog ■ *Occidozyga lima* SVL ♀ 5cm, ♂ 4cm
(*Bahasa Indonesia* Bancet Hijau Halus)

DESCRIPTION *Physique* Body small, plumped; head small with large, bulbous eyes located more dorsally than laterally on head; skin with scattered tubercules giving granular appearance to skin; series of small warts in lines on belly, flanks and chin; tympanum hidden by skin; no vomerine teeth; toes fully webbed; 2 distinct metacarpal tubercules; inner and outer metatarsal tubercule; large tubercule at back end of tarsus. *Colouration* Dorsum shades of brown, green or grey, sometimes mottled with lighter and darker patches; some individuals have thick or thin vertebral line of orange, light green or yellow along back; limbs weakly barred in some individuals; venter whitish, occasionally with brown spots. **TADPOLE** Free swimming; spiracle opening on side of body; snout pointed, thus distance between eyes 3–4 times distance between nostrils; mouth at end of snout (not just under tip as in most other species); lower lip horseshoe shaped; no labial teeth rows; brownish and well camouflaged against the bottom in water bodies. **HABITAT AND HABITS** Largely aquatic, and found in both slow-flowing and stagnant waters within grassland and forests, as well as plantations. Nocturnal, but regularly seen during day with nostrils and eyes exposed at the water's surface. Breeding call resembles bleating sound of a goat, *baa-baa-baa....* **NOTE** Vast genetic differences between nearly all populations of this 'species' have been discovered and it is likely that several lineages are grouped within this taxon.

Adult colouration.

Sumatran Puddle Frog ■ *Occidozyga sumatrana* SVL ♀ 5cm, ♂ 4cm
(*Bahasa Indonesia* Sumatera Bancet Hijau, Bancet Rawa Sumatera)

DESCRIPTION *Physique* Body small, plumped; head small with large, bulbous eyes located more dorsally than laterally on head; skin with scattered tubercules giving granular appearance to skin, especially at base of throat; tympanum hidden by skin; no vomerine teeth; toes fully webbed. *Colouration* Dorsum shades of brown, sometimes mottled with lighter and darker shades of browns, oranges or yellows; occasionally wide lighter stripe along back; venter whitish, occasionally with brown spots. **TADPOLE** Free swimming; spiracle opening on side of body; snout relatively broad, thus distance between eyes only

about twice distance between nostrils; mouth at end of snout (not just under tip as in most other species); lower lip horseshoe shaped; no labial teeth rows; tail long (more than twice body length) with narrow fins; dark spots on margin of fins and on body. **HABITAT AND HABITS** Inhabits puddles, ditches and slow-flowing streams in lowland forest and agricultural habitats, including paddy fields. Dorsally located eyes enable frog to stay in minimum exposure position at the water's surface. Males usually call (a short, squelching sound) from shallow waters, and small clutches of eggs are laid on wet ground close to water. Adults feed on insects, and tadpoles are carnivorous.

Adult colourations.

RANIDAE
TRUE FROGS

Formerly a large group, the 'old Ranidae' is now divided into several separate families. The 'new Ranidae' has been reduced to nearly 410 species. Ranids naturally occur worldwide except in large parts of southern South America and Australia.

Ranids have a generalized frog body plan with smooth and moist skin. The back limbs are large and powerful with extensive webbing between the toes. They range vastly in size from species smaller than a couple of centimetres, to the largest frog in the world, the Goliath Frog *Conraua goliath*, which reaches more than 30cm in snout to vent length, and weighs 3kg plus.

The vast majority of species are aquatic, so inhabit places close to water. A few species are semifossorial or arboreal. Some species living near noisy, rushing water bodies (such as waterfalls) use hand and foot waving as a method of communication instead of calling. Most species lay their eggs in the water and go through a typical tadpole stage.

Three species are known from Bali, of which the American Bullfrog *Aquarana catesbeiana* is introduced.

Although largely considered to be aquatic, some ranids, such as this White-lipped Frog, are often found on vegetation.

Cricket Frog ▪ *Amnirana nicobariensis* SVL ♀ 5cm, ♂ 4cm
(*Bahasa Indonesia* Katak Jangkrik, Kongkang Jangkrik)

DESCRIPTION *Physique* Body small; head narrow with pointed snout; distinct skin fold along each side of body; enlarged tubercules on flanks and back limbs; fingers and toes very long and with flattened discs; toes half webbed. *Colouration* Dorsum copper-brown, sometimes with very indistinct and slightly darker blotches; sides of head and body darker or chocolate-brown; upper lip whitish; venter cream or yellowish; back limbs barred with slightly darker shades. **TADPOLE** Free swimming; oval body; spiracle opening on side of body; 2 rows of papillae along upper lip; outer row of papillae along lower lip very long; labial teeth present; tail more than twice body length with wide fins; body light brown or orangey with darker brown mottling; tail patterned with shades of brown and black; margin of tail has black-and-white mottling. **HABITAT AND HABITS** Found in forested as well as man-made habitats, including those in cities. Breeds in permanent pools, rainwater ponds, tree holes and roadside ditches. Advertisement call is 6–10 repeated, sharp *kek-kek-kek-kek-kek...* notes, and males call from the ground or vegetation close to the water's edge.

Adult colouration.

American Bullfrog ■ *Aquarana catesbeiana* SVL 15cm
(*Bahasa Indonesia* Katak Banteng Amerika, Katak Lembu)

DESCRIPTION *Physique* Body large and robust; head narrow with pointed, oval snout; nostrils and eyes located more dorsally than laterally on head; tympanum large and distinct, and with supratympanic fold; skin wrinkled; toes fully webbed. *Colouration* Dorsum light brownish or greenish; occasionally with darker mottling on body, and darker markings and spots on limbs; venter whitish, occasionally with scattered darker mottling. **TADPOLE** Free swimming; oval body; spiracle opening on side of body; keratinized jaw sheaths and several rows of labial teeth with broad gap in second row on upper jaw; dorsum olive-green and speckled with black dots. **HABITAT AND HABITS** Introduced to Bali for restaurant meat and now found in several water bodies, mainly in midlands. Nocturnal and diurnal. Opportunistic predator that readily preys on smaller animals, including other frogs and conspecifics. Scavenging also known. Breeds in permanent water bodies with ample aquatic vegetation. Advertisement call a deep, loud, more or less metallic *ooom*. Males defend egg-laying sites voraciously, and females select a mate by entering his territory. Up to 20,000 eggs/clutch are laid, and reproduction may occur more than once a year.

Adults and juvenile (inset).

White-lipped Frog ■ *Chalcorana chalconota* SVL ♀ 7cm, ♂ 4.5cm
(*Bahasa Indonesia* Kongkang Kolam, Katak Berbibir Putih; *English* Schlegel's Frog)

DESCRIPTION *Physique* Body small (females significantly larger than males); head narrow with pointed snout projecting beyond mouth; tympanum distinct; skin of males covered with small tubercles; fingers and toes with flattened discs; toes fully webbed. *Colouration* Highly variable. Dorsum shades of green (mostly during day), brown or yellow (mostly at night); sides of body usually olive or brown; some individuals have darker spots or mottling; some have vague cross-bands on back limbs; inner sides of thighs reddish; upper lip white; sides of face usually darker than rest. **TADPOLE** Free swimming; oval body; spiracle opening on side of body; papillae only on corner of mouth; glandular patches on

body; tail about twice length of body, gradually tapering, acutely pointed; dorsum yellowish or straw coloured, with black spots or bars on body and face (sometimes as stripes radiating from the eye); patch of whitish glands on either side of ventral body brownish; tail transparent. **HABITATS AND HABITS** Usually inhabits quite vegetated habitats, especially those surrounding slow-flowing streams, but also found around paddy fields and ponds. Males call with complex combination of short, high-pitched chirping and whistling sounds (such as *twip*, *tik* or *tsp*) from edge of water or from vegetation. Clutches consist of more than 2,000 eggs, and tadpoles are usually found in deeper parts of pools.

Adult colourations.

RHACOPHORIDAE
ASIAN TREE FROGS

This large group is represented by about 425 species. It is ecologically related to the New World hylids in having a mostly arboreal lifestyle and a body shape adapted for this. Rhacophorids naturally occur in tropical sub-Saharan Africa and Asia.

Rhacophorids have enlarged toe discs at the ends of the fingers and toes to aid climbing. Most species have large eyes with horizontal pupils. While the dorsum colouration is usually plain, many have bright, flashy and patterned colourations on the inner thighs. When the frog leaps, this colour pattern distorts its overall body pattern and confuses predators. A few Southeast Asian members have extensive webbing, allowing them a limited gliding capacity.

Some species make foam nests in vegetation above water by beating their limbs while in amplexus. The eggs are laid in the foam, then covered with seminal fluid. They hatch within the foam nest and the tadpoles then fall into the water below. Other species lay eggs on the damp forest floor and develop directly.

One rhacophorid species is known from Bali.

Arched backs and long digits with discs at the end are characteristics of rhacophorids, such as this Common Southeast Asian Tree Frog.

Common Southeast Asian Tree Frog

■ *Polypedates leucomystax* SVL ♀ 8cm, ♂ 5cm
(*Bahasa Indonesia* Emlegan, Katak Pohon Bergaris, Katak Pohon Coklat)

DESCRIPTION *Physique* Body medium in size, slightly built; snout pointed; lacks raised warts or tubercules on back; finger and toes with noticeable circular discs; very little webbing between fingers; toes fully webbed except on toe 4. *Colouration* Dorsum varies among shades of brown and green-grey; can be either patternless, with darker irregular spots, or with up to 6 darker stripes running along back; lateral face darker than dorsal, with prominent dark stripe from snout through eye and above tympanum to past neck; limbs have dark cross-bars; usually lighter in colour at night than while resting in diurnal refuge. **TADPOLE** Free swimming; oval body; eyes set on sides of body (not towards top); spiracle opening on side of body; papillae only on corners of mouth; glandular patches on body; tail widest in middle and with pointed tip; body light brown and variegated, and mottled with white, yellow, darker shades of brown, green and black; or orangey with darker brown mottling; venter whitish. **HABITAT AND HABITS** Very adaptable and opportunistic species inhabiting most natural and man-made habitats. Predominantly arboreal and nocturnal. Adults feed on arthropods, as well as small vertebrates (like geckos). Breeding takes place year around in wetter parts of Bali, and with the onset of rains in drier parts. Call a much-spaced, nasal quack as from a duck. Up to 400 eggs deposited in oval-shaped foam nests attached to twigs, leaves or walls overhanging stagnant water. Tadpoles are voracious feeders and also display cannibalism.

Adult colourations.

EMYDIDAE
New World Pond Turtles

With about 55 species included, this family is native to the western hemisphere.

The carapace is hard, domed or low arched, with 11 pairs of sutured peripherals around the margin. Some species have keels. The plastron is often large and without mesoplastron scutes. The toes are usually webbed and the neck is drawn back vertically into the shell.

Most members of this family inhabit fresh water, while a few occur in brackish water. They are generally omnivorous, but some species shift from a carnivorous diet as juveniles to a herbivorous one as adults. Males display courtship rituals, vibrating their long claws at females. Multiple clutching may occur within a year.

The Red-eared Slider *Trachemys scripta*, native to the Mississippi River drainage in the USA, is known from several introduced populations in Bali, especially in ponds in urban areas. The Asian Turtle Conservation Network has listed naturalized populations of this species from several other Indonesian islands. It is also commonly sold as a pet in some parts of Indonesia.

Juvenile Red-eared Sliders are commonly sold as pets in parts of Asia.

Red-eared Slider ■ *Trachemys scripta elegans* Shell 25cm
(*Bahasa Indonesia* Kura Telinga Merah, Kura Brazil)

DESCRIPTION *Physique* Round and domed carapace with mostly smooth edge; plastron without movable hinge; males have large claws on fingers and longer tails than females.

Colouration Carapace greenish, olive or brownish, with or without yellowish lines or a central spot; plastron yellow with most scutes having round black blotch; head and neck distinctly striped in yellow; distinct orange or red elongated spot behind eye; juveniles bright green with yellow stripes and bars on limbs, head and carapace.
HABITAT AND HABITS Diurnal and commonly seen basking in open places, and also floating on the surface of water. Catholic in diet, feeding on vegetation as well as an array of animals. Oviparous, with up to 25 eggs in each clutch. Known to be a carrier of *Salmonella* pathogens.

Adult colourations.

GEOEMYDIDAE
ASIAN HARD-SHELLED TURTLES

This family naturally occurs in the tropics and subtropics of Asia, North Africa, Central and South America, and Europe. Over 70 species are included in this family.

Members of the group have an oblong to oval, domed or depressed carapace with 24 marginal scutes. The plastron has 12 scutes but no mesoplastron. The toes are usually webbed, and the neck is drawn back vertically into the shell.

Geoemydids generally inhabit fresh water; a few species occupy coastal marine environments and terrestrial habitats. The vast majority of species are herbivorous or omnivorous, but juveniles of most are largely carnivorous. Scavenging is a common feeding habit. They are oviparous and have small clutches.

Three members of the family are known from Bali.

Geoemydids such as this Southeast Asian Box Turtle can draw back their necks vertically into the shell.

Southeast Asian Box Turtle ■ *Cuora amboinensis* Shell 20cm
(*Bahasa Indonesia* Kura Ambon, Kura Batok)

DESCRIPTION *Physique* Oval, highly domed carapace with no anal notch; smooth surface with single weak vertebral keel in adults and 3 (1 on either side of the vertebral one) in juveniles; 5 vertebrals; plastron has movable hinge at all life stages, which enables shell to be completely closed; upper jaw weakly hooked; feet webbed. *Colouration* Carapace black, dark brown or olive-brown; plastron cream or yellow, with each scute having a single, large, irregular black blotch; face and dorsal side of neck dark, with bright longitudinal yellow stripes on lateral sides; throat cream or whitish. **HABITAT AND HABITS** Known from West Bali National Park area, where it is said to occur in aquatic habitats in monsoon forests and on agricultural land. Occupies both slow-flowing and stagnant waters. Adults show terrestrial behaviours, while juveniles are prominently aquatic. Omnivorous, feeding on aquatic plants, invertebrates (including earthworms) and also fungi. Oviparous, with clutch size of up to 6 eggs. Common in pet industry, so escapees or released individuals may also be encountered.

Adult and close-up of head (inset).

Asian Leaf Turtle ■ *Cyclemys dentata* Shell 25cm
(*Bahasa Indonesia* Kura-kura Daun)

DESCRIPTION *Physique* Oval, depressed carapace with 1 vertebral and 2 lateral keels, and 5 vertebral scutes; plastron with fine notch at back end and movable hinge in adults; forehead with enlarged scales. *Colouration* Carapace dark brown, mostly unpatterned but occasionally with fine black radiating lines; plastron light brown or yellowish, with radiating dark lines; plastron of juveniles mottled with black; head brown with no distinct markings, or with pale orange or orange-brown lines on sides (more prominent in juveniles). **HABITAT AND HABITS** Occupies slow-flowing and stagnant waters, including on agricultural land. Adults show terrestrial behaviours, while juveniles are predominately aquatic. More or less walks on the bottom of water bodies rather than swimming. Crepuscular (active at dawn and dusk). Omnivorous, feeding on invertebrates, plant matter (fruits) and carrion. Hole nester, laying up to 4 elongated, hard-shelled eggs.

Adults and juvenile (top right).

Black Marsh Turtle ■ *Siebenrockiella crassicollis* Shell 20cm
(*Bahasa Indonesia* Kura Pipi Putih)

DESCRIPTION *Physique* Oval, carapace with 1 prominent vertebral keel in adults, and 2 additional lateral keels in juveniles; edge of carapace strongly serrated towards the back; plastron with shallow 'U'-shaped notch at back end and not hinged; both front and hind limbs webbed; neck characteristically thick. *Colouration* Carapace almost entirely black or dark grey and unpatterned; plastron entirely black, dark brown or yellowish with splotches or patterns of darker colours; head brown or dark grey with light markings around eyes and throat in juveniles and adult females, but fade with growth in males. **HABITAT AND HABITS** Largely aquatic. Occupies slow-flowing or stagnant waters, including marshes, ponds, lakes and agricultural land. Nocturnal and omnivorous, feeding on aquatic animals as well as rotting plants and fruits. Also known to scavenge on carcasses of larger animals. Lays 3–4 clutches a year, each with 1–2 relatively large eggs. **NOTE** The recent records of individuals from Bali could be escaped pets.

Adult female colouration.

TRIONYCHIDAE
SOFT-SHELLED TURTLES

Thirty-three members of this family naturally occurs in Asia, Africa and North America.

With a few exceptions, the carapace of most species is flattened, skin clad and lacks horny scutes. The central part of the carapace is more rigid, while the edges are more leathery. The nostrils are located on a short trunk, or proboscis, and the neck is disproportionately long compared with the body size. The feet are webbed, with three claws on each limb. These are among the largest freshwater turtles in the world, and females are substantially bigger than males.

Soft-shelled turtles predominantly inhabit fresh water, and their 'soft' bodies allow them to move quite flexibly and fast in water. They use their long necks and snorkel-like nostrils to breathe air at the surface, while the body remains submerged underwater. They are also capable of 'breathing' underwater through the highly vascular skin layer in the mouth (similar in function to gill filaments in fish), allowing lengthy underwater stays. Most species are carnivorous, and many have powerful crushing processes on the outer borders of their mandibles.

One species is known from Bali, and the Malayan Softshell Turtle *Dogania subplana* could possibly occur there.

The nostrils of soft-shelled turtles are characteristically located on a short trunk, or proboscis.

Southeast Asian Soft Terrapin ■ *Amyda cartilaginea* Shell 60cm
(*Bahasa Indonesia* Labi-labi, Bulus)

DESCRIPTION *Physique* Carapace flattened and slightly oval, with round sides and distinct raised tubercules on front margin (as opposed to straight sides without tubercules in the Malayan Softshell Turtle *Dogania subplana*, possibly occurring in Bali); head narrow with tubular nose; digits strongly webbed, with large claws on 3 digits. *Colouration* Dorsum brown, purplish-black, olive-grey or greenish-grey, and largely unpatterned in adults but with yellow-bordered black streaks or spots (usually 4 large spots) in subadults and juveniles; plastron colour shows sexual dimorphism, with grey in females and white or cream in males; head usually marked with yellow spots. **HABITAT AND HABITS** Highly aquatic, and inhabits both stagnant and flowing waters. Nocturnal. Mostly carnivorous, feeding on invertebrates, amphibians and fish, but also occasionally taking plant material and carrion. Digs holes in riverbanks and lays up to 40 eggs. Multiple clutches may be laid in a year.

Adult and juvenile (inset).

CHELONIIDAE
HARD-SHELLED SEA TURTLES

Six species of truly marine turtles comprise this family. Most members have a worldwide distribution in tropical and temperate waters.

The carapace is smooth, large and keratinized with strong, horny scutes. The head has keratinized epidermal scutes, and can be partially retracted under the carapace. A well-developed, horny beak, or 'tomium', covers both mandibles. The front flippers are large and paddle-like, while the back ones are more rounded and shorter. Both pairs are non-retractable, and at least with one claw on each. Males can be distinguished from females by their longer tails and larger claws.

These turtles are highly aquatic. Only females generally come ashore, to lay eggs, sometimes on communal nesting beaches. In a few parts of the world the males may come ashore to bask. Females usually show nesting site fidelity, and return to lay eggs near the spot where they left the last clutch, or even on the same beach from which they emerged as hatchlings. Nesting takes place on sandy beaches, and clutches of over 100 eggs are buried in the sand and left unattended. These turtles display temperature-dependent sex determination (TSD). The hatchlings lead a pelagic-nectonic life, but very little is known of the juvenile stages. The diet varies with the species, and ranges from seaweeds to corals and jellyfish.

Four species nest in Bali, while the Flatback Turtle *Natator depressus* possibly occurs in its waters.

Hard-shelled sea turtles are integral parts of coral reef ecosystems in the tropics.

Loggerhead Turtle ■ *Caretta caretta* Shell 110cm
(*Bahasa Indonesia* Penyu Tempayan)

DESCRIPTION *Physique* Carapace heart shaped with juxtaposed scutes; 5 centrals, 5 pairs of laterals, and commonly 12 or 13 pairs of marginals; head large, subtriangular and broad, with 2 pairs of prefrontal scales, and a very strong, horny beak; front flippers relatively short and thick, with 2 claws; back flippers with 2 or 3 claws; hatchlings and juveniles with 3 longitudinal keels made from blunt spines on carapace. *Colouration* Carapace reddish-brown or dark brown; plastron yellow-creamy; carapace of hatchlings dark brown, while flippers are pale brown marginally. **HABITAT AND HABITS** Adults primarily inhabit continental shores of warm tropical and subtropical seas. They usually nest solitarily on communal nesting beaches. Females lay multiple clutches, each with 25–190 eggs. Predominantly carnivorous at all stages of life, feeding on a wide array of marine organisms.

Adult (top left), *subadult* (top right), *hatchling* (bottom left), *and arrangement of prefrontals and dorsal scutes* (bottom right).

Green Turtle ■ *Chelonia mydas* Shell 150cm
(*Bahasa Indonesia* Penyu Hijau)

DESCRIPTION *Physique* Carapace more or less oval with juxtaposed scutes; 5 centrals, 4 pairs of laterals, and commonly 12 pairs of marginals; juveniles with low keels on centrals; head with 1 pair of elongated prefrontals and weak beak; front flippers long and with 1 claw; back flippers short and with 1 claw. *Colouration* The English name Green Turtle is derived from the colour of turtle fat, which was once highly sought after for turtle soup. Carapace highly variable, ranging from plain brownish or greenish colours to vivid combinations of brown, yellow and greenish tones, forming radiated bands, or numerous dark blotches; plastron pale yellow; head and flipper scales of juveniles edged with yellow or white; newborn hatchlings dark brown or nearly black on dorsum and white on venter.

HABITAT AND HABITS Inhabits tropics and common around oceanic islands. Usually solitary and nektonic, but occasionally forms feeding aggregations in shallow-water areas with abundant seagrass or algae. Carnivorous in juvenile stage, but becomes strictly herbivorous as adult, only consuming seagrass and seaweeds. Clutch size 40–200 eggs.

Adult (top left), *subadult* (top right), *hatchling* (bottom left), *and arrangement of prefrontals and dorsal scutes* (bottom right).

Hawksbill Turtle ■ *Eretmochelys imbricata* Shell 100cm
(*Bahasa Indonesia* Penyu Sisik)

DESCRIPTION *Physique* Carapace heart shaped or elliptical, with imbricated scutes in adults; 5 centrals, 4 pairs of laterals, and 11 pairs of marginals plus one pair of postcentral scutes; juveniles with 3 keels of spines along carapace; head medium sized, narrow and with pointed, bird-like beak; 2 pairs of prefrontal and 3–4 postorbital scales; tomium hooked at tip and not serrated on cutting edge; each flipper has 2 claws. *Colouration* Carapace highly variable, but generally olive-brown with spots or stripes (usually arranged in fan-like pattern) of brown, black, red and yellow; head and flipper scales usually black and have thick yellowish borders; newborn hatchlings dark brown or nearly black on dorsum, and white on venter. **HABITAT AND HABITS** Inhabits littoral waters and feeds (sometime communally) in reefs, lagoons and estuaries. Carnivorous, feeding on animals that inhabit hard substrates such as corals, tunicates, shellfish and sponges, including toxic species. Toxic chemical compounds from prey accumulate in the turtles' tissues, so consumption of their meat may cause serious illness and in extreme cases even death in humans. Usually nests solitarily, with several clutches – each comprising up to about 250 eggs – being laid within a year.

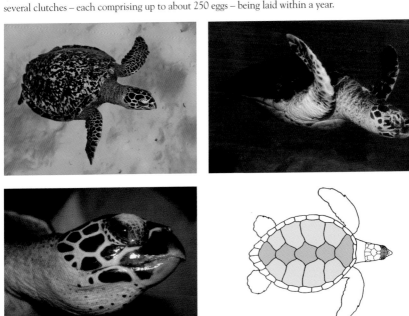

Adults (top), *beak-like head* (bottom left), *and arrangement of prefrontals and dorsal scutes* (bottom right).

Olive Ridley Turtle ■ *Lepidochelys olivacea* Shell 120cm
(*Bahasa Indonesia* Penyu Lekang)

DESCRIPTION *Physique* Carapace heart shaped or near round, with flattened top; males have more tapered shells than females; scutes juxtaposed; 5 centrals, 5–9 (usually 6–8) pairs of laterals, and 12 pairs of marginal scutes; posterior marginals serrated; hatchling shell has 3 rows of keels; head medium in size, subtriangular; 2 pairs of prefrontal scales; front flippers with 1–2 claws, and sometimes another small claw on distal part; back flippers have 2 claws. *Colouration* Carapace olive-grey or brown; plastron creamy or whitish, with pale grey margins; hatchlings dark grey dorsally, and cream or white underneath.

HABITAT AND HABITS Adults occur in shallow waters and coastal habitats such as mangroves, while some juveniles occupy the open ocean. Usually sleeps floating on the surface at night. Predominantly carnivorous and feeds on molluscs, crustaceans and jellyfish. Females sometimes nest in groups of thousands of individuals, commonly referred to as 'arribada'. Each lays clutches of up to 160 eggs.

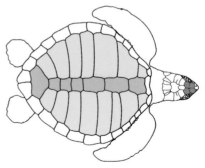

Adults (top), hatchling (bottom left), and arrangement of prefrontals and dorsal scutes (bottom right).

DERMOCHELYIDAE
LEATHERBACK SEA TURTLES

A single truly marine turtle species comprises this family.

The carapace is strongly rigid and without keratinized epidermal scutes, except in hatchlings. The dermal bones forming the shell are absent, and are replaced by a mosaic of small, polygonal platelets. The front flippers are very large and paddle-like, with no claws.

The sole member of this family regularly migrates into cold Arctic waters, probably to feed on jellyfish. Females mature at 9–14 years. This is one of the few chelonians that are known to communicate vocally. The adults produce different types of sound when out of the water (sometimes in response to distress) and hatchlings make sounds even while inside the nest, presumably to coordinate group behaviour.

Leatherback Turtles nest infrequently on southern beaches of Bali.

Occasionally reaching 3m in length, the Leatherback Turtle is the largest turtle in the world.

Leatherback Turtle ■ *Dermochelys coriacea* Shell 200cm
(*Bahasa Indonesia* Penyu Belimbing)

DESCRIPTION *Physique* Carapace elongated and tapering towards rear; 7 dorsal and 5 ventral longitudinal keels; adults have rubber-like, leathery skin, but juveniles have body scales; head medium in size, subtriangular; 2 pairs of prefrontal scales; front flippers very large (exceeding half of carapace length in adults), paddle shaped and without claws. *Colouration* Dorsum black sprinkled with white blotches, usually along keels; whitish or pinkish blotches on neck, shoulders and groin; females with distinct pink patch on forehead. **HABITAT AND HABITS** Occurs in open oceans. The deepest diving reptile, capable of reaching depths of more than 1,000m, presumably in search of zooplankton. Carnivorous, and primarily feeds on large zooplanktonic invertebrates such as jellyfish and tunicates, as well as crustaceans, fish and seaweeds. Mouth and oesophagus lined with long, backwards-projecting spines that assist the turtles to swallow slippery prey such as jellyfish. Usually nests solitarily and lays multiple clutches, each comprising up to about 160 eggs.

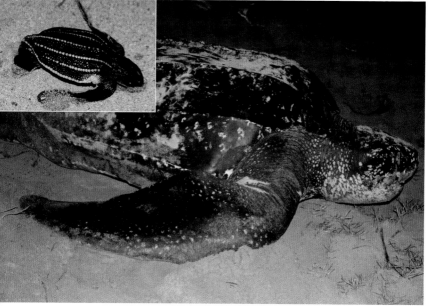

Adult and hatchling (inset).

AGAMIDAE
AGAMID LIZARDS/DRAGONS

More than 505 members of this family are found in the Old World, in Asia, Africa, Europe and Australia.

The bodies of agamid lizards are covered with imbricated scales lacking osteoderms; the limbs are well developed, and the back limbs are usually longer than the front ones. The tail is long and prehensile in some species; it does not usually show autotomy, but regeneration occurs in some species if the tail is broken. Many species have body ornamentation in the form of rostral appendages, dewlaps, crests or patagia. Femoral and preanal pores are present in most species. These lizards are closely related to chameleons and iguanas, and are differentiated by skeletal and dentition features.

Agamids inhabit areas ranging from coastal sand dunes to cloud forests in the highest mountains. They are diurnal, with most species sleeping on the tips of branches on tree trunks at night. Others sleep in crevices, burrows or even in water. Southeast Asian agamids are generally arboreal or semiarboreal, and a few are terrestrial. Most species are able to escape fast from danger, while a few rely on cryptic colouration and shape to avoid predation. They are capable of relatively rapid colour change, and generally have complex visual communication behaviour. The majority of species are oviparous, producing eggs with leathery shells, but a few are ovoviviparous. They are predominantly carnivorous, though some larger species are omnivorous.

Four species of agamids are known from Bali. A specimen of a Lined Gliding Lizard *Draco lineatus* from Bali now deposited at the Western Australian Museum is likely to be an erroneous one, as the species is only known from the northern Maluku and Maluku island groups in eastern Indonesia. However, additional species, such as the Red-barbed Flying Lizard *D. haematopogon*, may have been historically present in Bali.

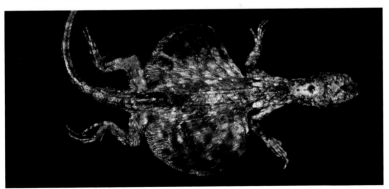

Lined Gliding Lizaed supposedly collected fom Candikuning in Bali (preserved specimen WAM.R109102).

Great Crested Canopy Lizard ■ *Bronchocela jubata* SVL 13cm, TL 55cm
(*Bahasa Indonesia* Bunglon Surai; *English* Maned Forest Lizard)

DESCRIPTION *Physique* Body medium, slender and compressed; head large, with large gular pouch extending under front limbs; tail long, thin and pronouncedly keeled at thicker base. *Scales* Supralabials 9–10; infralabials 8–9; crest on neck high, with sickle-shaped scales (especially in males), facing backwards; crest on dorsal body shorter, like a

serrated ridge, extending to tail (does not extend to tail in the Green Crested Lizard B. *cristatella*, which some have claimed to be present in Bali); dorsals and ventrals keeled; 2 larger keeled scales on front border of tympanum. *Colouration* Dorsum bluish-green, yellowish-green or brownish-green, changeable to darker and duller shades; usually has white stripe beneath tympanum; sometimes lighter spots or vertical bars with darker edges on body and shoulders (dorsum is darker when cold, showing markings more clearly); some individuals have bright red, blue or yellow markings; venter pale.
HABITAT AND HABITS Inhabits vegetation in forests, scrubland and home gardens. Arboreal and diurnal, usually sleeping on thin vegetation at night. Feeds on arthropods. Oviparous, laying 2 spindle-shaped eggs, which are buried in sandy soil beneath a layer of humus under bushes.

Subadult (top) and adult male (bottom)

Common Garden Lizard ■ *Calotes versicolor* SVL 10cm, TL 35cm
(*Bahasa Indonesia* Bunglon Taman; *English* Common Bloodsucker)

DESCRIPTION *Physique* Body medium, compressed; head large, especially in adult males with small gular sac on throat. *Scales* 2 small groups of spines completely separated from each other above each tympanum; crest on neck high (especially in adult males), shorter on body, extending to base of tail in large individuals; scales on body facing backwards and upwards; scales on underside of thigh keeled. *Colouration* Dorsum generally light brown but extremely diverse and changeable; generally broad brown bands across the back, interrupted by a yellowish lateral band; black streaks radiate from the eye; breeding males with red heads and necks, and black shoulder patch; most females with 2 longitudinal, light coloured lines from eye to tail; juveniles light brown with dark transverse lines on body and tail. **HABITAT AND HABITS** Inhabits cities, farmlands and open natural habitats. Arboreal and diurnal, usually seen on low vegetation, tree trunks and fences. Feeds mainly on arthropods, but also takes plant parts and small vertebrates. Also shows cannibalism. Males are territorial and display from an elevated site within their territory doing 'press-ups' and head nodding. Oviparous, laying 6–20 soft leathery eggs.

Male (top) Juvenile (bottom left) and Female (bottom right)

Fringed Flying Lizard ■ *Draco fimbriatus* SVL 5cm, TL 9.5cm
(*Bahasa Indonesia* Cecak Terbang Fimbriata)

DESCRIPTION *Physique* Body small, slender, depressed; head medium with depressed snout; eyes large; lower eyelid fused with rudimentary upper eyelid to form fixed spectacle that is surrounded by granular scales; hind limbs about as long as distance between limbs; 19–24 subdigital lamellae under 4th toe; gliding membrane (patagium) supported by 5 ribs. *Scales* No postnasal; frontal almost equal in size to prefrontals that are usually in broad contact; supralabials 7; infralabials 5–7, supraciliaries 5–7, but usually 6; small, thorn-like scale on supraciliary edge; body scales in 24–28 rows around midbody; male has crest on tail consisting of compressed triangular scales. *Colouration* Dorsum light brown or golden-brown, with blackish central line from snout to base of tail; each side of body has dark band commencing from eye and backwards, which broadens at body where it has lighter speckles; ventral parts of patagium yellowish-grey without any spots; gular pouch of male white with yellow tip, and bright blue with orange-yellow tip in female; limbs dark and highly speckled with pale spots; tail brownish or bronzy above, and darker on sides; venter white in male and reddish in female. **HABITAT AND HABITS** Known from a historical record from Tjandikesuma area in West Bali. Arboreal, moving actively on tree trunks; very rarely found on the ground. Diurnal, and feeds on ants. Oviparous.
NOTE The subspecies *hennigi* recorded from Bali is under taxonomic review.

Adult and ventral side of gliding membrane (inset).

Common Flying Lizard ▪ *Draco volans* SVL 7cm, TL 13cm
(*Bahasa Indonesia* Haphap, Cekiber, Dangap-dangap, Cecak Terbang)

DESCRIPTION *Physique* Body small, slender; female noticeably bigger than male; head small; subtriangular dewlap with slightly enlarged scales; gliding membrane (patagium) supported by 6 (very rarely 5 or 7) ribs; hind limbs shorter than distance between limbs. *Scales* Supralabials 6–12; crest on neck short with triangular scales in both sexes; large, thorn-like scale on supraciliary edge; dorsals smooth or keeled, unequal in size; no crest on tail. *Colouration* Dorsum greyish, brownish or yellowish-tan, with darker and lighter marbling; head, neck and vertebral line generally has black spots; patagium pale yellow-tan or yellow-brown in ground colour, highly variable in colour and pattern of markings, but male has distinct dark bands near outer edge, and female has reticulated pattern at edge; dewlap yellowish or light orangey, with small dark spots at base in male, bluish-grey in female; venter whitish or bluish-grey. **HABITAT AND HABITS** Inhabits vegetation in lowland and submontane forests, as well as urban parks and home gardens. Arboreal, generally occupying tree trunks, and gliding from tree to tree. Diurnal and territorial. Utilizes complex visual displays involving extended dewlaps, patagia, head bobbing and circling another individual. Feeds on insects and possibly other smaller arthropods. Oviparous, with clutch size of up to 6 eggs buried in soil.

Adult and dorsal side of gliding membrane (inset).

CHAMAELEONIDAE
CHAMELEONS

Over 210 species of chameleons are currently known. Apart from four species known from the Mediterranean, the Arabian Peninsula and South Asia, all other species naturally occur in sub-Saharan Africa and Madagascar.

Chameleons vary greatly in body form and size, ranging from the 1.5cm long *Brookesia micra* to the nearly 70cm long Malagasy Giant Chameleon *Furcifer oustaleti*. Most have ornamental structures such as horns, nasal appendages and crests on their head or face. Some are sexually dimorphic, with the males being more ornamented and sometimes significantly different in colour. Their feet have opposable toes, giving them a tongs-like appearance, to firmly grip branches. Tails are prehensile. Their distinctive 'turret' eyes that move independently of each other provide 360 degree vision. They primarily feed on insects by whipping out their tongues to capture prey located some distance away.

There are anecdotal reports of chameleons from forests at the Bedugul mountain area in central Bali. These are of deliberate and accidental introductions. There are also recent reports of locals selling chameleons along the roadside in Bedugul. The author or his colleagues have not seen any specimens from Bedugul. However, what is most certainly an escaped adult belonging to the Short-horned Chameleon complex (either *Calumma brevicorne* or *Calumma crypticum*) was recorded from a grassland habitat atypical for these chameleons north of Lake Palasari in western Bali in 2015.

Short-horned Chameleon from western Bali.

GEKKONIDAE
COSMOPOLITAN GECKOS

With more than 1,240 members, this is one of the largest reptile families. Geckos are near global in distribution, occupying all continents (except Antarctica), as well as most isolated islets in the world. Some species easily adapt to human habitats, a characteristic that assists their dispersal.

The body of a gecko is covered with velvet-like, soft skin containing non-glossy, usually juxtaposed scales. Some species have tubercules. Geckos lack eyelids and use the broad, flat tongue to clean the large eyes, which are covered with clear spectacles. The fingers and toes range from bent and bird-like to flat and well padded, most with numerous microscopic setae that allow them to climb smooth surfaces. This is one of the few vocal reptile groups, and some species use sound to communicate.

Geckos live in terrestrial and arboreal habitats. Tail autotomy is prominent in the group. Most members are nocturnal. They primarily feed on arthropods, but larger species also take small vertebrate prey. They are also known to lap soft fruits or sap. The clutch size is normally two, but multiple clutching as well as communal egg laying occurs in some members of the family. Unlike most other reptile groups that have soft-shelled eggs, gekkonids have calcified, hard-shelled and brittle eggs.

At least seven species have confirmed records from the island, and several others are likely to occur there. Records of the Indo-Pacific Gecko *Hemidactylus garnotii* from Bali need further verification.

Many species of gecko inhabit human-made habitats, where they can reach high densities.

Bent-toed Gecko ▪ *Cyrtodactylus* spp. SVL 5–8cm, TL 13–15cm
(*Bahasa Indonesia* Cicak Tanah, Token Pohon Asam)

DESCRIPTION *Physique* Body moderate or large, and more or less stout; head large; snout elongated; fingers and toes bent at an angle, slender, not dilated or slightly dilated; pupil vertical; tail long and normally longer than head and body; preanal grove and preanofemoral pores present in males. *Scales* Dorsal surface has granular scales mixed with scattered, rounded, smooth or weakly keeled tubercules. *Colouration* Dorsum usually light brown, greyish-brown or purplish-brown, with darker and sometimes white spots that may form irregular transverse bars; some individuals have large and dark blotches on body; tail usually dark banded; venter cream or brown, unpatterned or with black spots. **HABITAT AND HABITS** Occupies natural and man-made habitats from coasts to submontane forests in mid-hills. Mostly arboreal, but also inhabits rocks, boulders, walls, abandoned houses and similar places. Animals inhabiting houses sometimes move on to surrounding vegetation at night-time. Sometimes employs caudal luring before attacking prey. Clutches of 2 eggs usually laid under bark or in crevice. **NOTE** Historically, researchers have assigned the 'species' found in Bali to either the **Clouded Bent-toed Gecko** *Cyrtodactylus marmoratus* or the **Tamarind Bent-toed Gecko** *C. fumosus*. However, recent molecular, morphological and museum studies have revealed that the bent-toad geckos in Bali represent neither of these species, and possibly comprise several species that are new to science.

Hatchling (bottom right) *and adults.*

Four-clawed Gecko ■ *Gehyra mutilata* SVL 5cm, TL 12cm
(*Bahasa Indonesia* Cecak Gula)

DESCRIPTION *Physique* Body stout and depressed, with lateral fold; head relatively large; tail smooth, carrot shaped and with large, flat scales on venter; fingers and toes widened; inner finger and toe frequently without visible claw (rarely with minute one); distinct web connects thigh and calf in adults; males have 25–41 preanofemoral pores. *Colouration* Dorsum light brownish, yellowish or cream, with or without two longitudinal rows of pale spots along back and irregular darker spots scattered on body; sides of face have darker band from snout to tympanum, and indistinct white band; juveniles darker than adults, with bright white and darker spots making lines along body. **HABITAT AND HABITS** More commonly found in human habitats than in natural forests. Has 'loose' skin that sometimes gives a dead appearance and easily comes off when roughly handled (often interpreted as 'self-mutilation'). Able to change body colour from light to dark and from spotted to plain with considerable rapidity. Nocturnal, and larger individuals seem territorial. Apart from small arthropods, also feeds on nectar and fruit juices. Lays 2 hard-shelled eggs, which are fused together terminally.

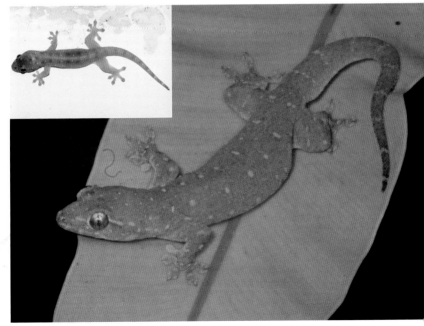

Adult and hatchling (inset).

Tokay Gecko ▪ *Gekko gecko* SVL 15cm, TL 25cm
(*Bahasa Indonesia* Tokek)

DESCRIPTION *Physique* Body large, robust and depressed; head large; lamellae under 4th toe 20–23. *Scales* Subconical tubercules among granular scales on body and tail, usually arranged as rows; supralabials 12–14; infralabials 10–12; males with 10–24 preanal pores; femoral pores also present. *Colouration* Dorsum of adults slaty-grey or bluish, with striking orange dorsal patterning; juveniles have cream spots, sometimes merging into pale bands; tail dark banded, with banding most prominent in juveniles; venter cream, unpatterned or spotted with pink. **HABITAT AND HABITS** Commonly associated with human-modified environments (including insides of buildings), forests, plantations (especially coconut trees) and cave systems. Mostly nocturnal, but sometimes seen in ambush pose by day. Displays territorial behaviours, but sometimes occurs in communal groups. Distinctive call, *tok-keh*, by males is supposed to be the source of the common name. Shows aggressive, open-mouthed displays when threatened. Feeds on wide variety of prey ranging from arthropods, lizards and small mammals to snakes. Oviparous, with clutch size of 1–3 eggs.

Hatchling (top left) *and adults.*

Asian House Gecko ■ *Hemidactylus frenatus* SVL 5cm, TL 12cm
(*Bahasa Indonesia* Cecak Rumah, Cecak Kayu)

DESCRIPTION *Physique* Body medium in size and slightly depressed, without cutaneous folds on sides but scattered round tubercules among granular scales; body mostly smooth; tail slightly depressed, segmented, with rows of enlarged tubercules and small spines on sides of original tail (not on regrown tails); no webbing in fingers and toes; 8–11 lamellae under 4th toe undivided, subsequent 5–8 divided; males have continuous series of 15–36 preanofemoral pores. *Colouration* Dorsum greyish-brown or dusky brown, usually uniform but sometimes with indistinct, wavy dark longitudinal lines enclosing pale dashes or spots; brown band (varying intensities) from nostril to above neck, bounded by darker lines, may be present or absent; venter unpatterned cream or light beige; juveniles darker and more prominently patterned than adults. **HABITAT AND HABITS** Commonly found in human habitats as well as natural forests. Nocturnal, but feeding during daytime can be seen. Site fidelity is often observed, and sometimes many individuals share the same feeding spots around artificial lights. Commonly feeds on rice, bread and other man-made food, so has become a 'pest' in some houses. Call consists of a series of 4–5 loud stacatto notes, *chic chic chic…* Usually lays 2 eggs, which are attached to substrate, but communal egg laying is also known.

Adult and hatchling (inset).

Flat-tailed Gecko ■ *Hemidactylus platyurus* SVL 6cm, TL 12cm
(*Bahasa Indonesia* Cecak Berekor Pipih, Cecak Rumah)

DESCRIPTION *Physique* Body medium in size and depressed, smooth and without keeled or imbricate scales; head with long snout; tail strongly flattened, with fringe of soft and spiny skin along sides of body; noticeable cutaneous skin webbing on thighs; fingers and toes have prominent webs; 6–9 lamellae under 4th toe undivided, subsequent 5–7 divided; males have 34–36 femoral pores. *Scales* Supralabials 9–11; infralabials 7–8. *Colouration* Dorsum ranges from light greyish to brownish-brown or dusky-brown, sometimes marbled with elongated darker spots or variegation; usually dark streak from eye to shoulder; tends to be lighter in colour during day or when well warmed up. **HABITAT AND HABITS** Commonly found in human habitats as well as natural forests. A large number of individuals may co-occupy a given house or space. Nocturnal and crepuscular. Well adapted to climbing. Feeds on small arthropods and occasionally eggs of geckos. Also commonly feeds on rice, bread and other man-made foods, so has become a 'pest' in some houses. Utters several types of vocalization, ranging from squeaks to a *tchick tchick tchickickick* to a quick *djieb-djieb-djieb*. Oviparous, laying 2 hard-shelled eggs. Occasional communal egg laying is known.

Cutaneous skin webbing on adult (top right) *and adults.*

Common Dwarf Gecko ▪ *Hemiphyllodactylus typus* SVL 4.5cm, TL 9cm
(*Bahasa Indonesia* Tokek Cebol)

DESCRIPTION *Physique* Body slender, elongated; head small and only slightly distinct from neck; first inner toe stunted, clawless or with minute claw; 3–6 lamellae under 4th toe part divided, part undivided; tail long, slender and prehensile; populations of this gecko are unisexual and only consist of females which, however, possess 8–10 femoral pores and 10–12 preanal pores. *Scales* Supralabials 10–14; infralabials 10–11. *Colouration* Dorsum dark brown or yellowish-brown, with some individuals having darker chevrons or dorsolateral lines of large black marks; dark brown stripe from nostril to shoulder; tail generally more pale, with characteristic dark-edged mark at base; venter cream with dark brown speckles; juveniles more purple in colour than adults, with pink spots on face, body and feet, and more orangey tail. **HABITAT AND HABITS** Inhabits both man-made and forested areas, including mangroves. In the wild, individuals stay high up on trees and descend during the night for feeding. Adults generally slow moving, but juveniles are fast and can jump considerable distances. Feeds on insects and spiders. This all-female unisexual species reproduces by parthenogenesis. Lays 2 eggs under bark or in a crevice, though eggs are occasionally found attached to open trunks and to leaf axils.

Adult and structure of digits (inset).

Lombok Mourning Gecko ▪ *Lepidodactylus lombocensis* SVL 4.5cm, TL 9cm
(*Bahasa Indonesia* Tokek Duka)

DESCRIPTION *Physique* Body elongated, slender; head longer than broad; dorsals granular and without tubercules, 110–112 rows around midbody; tail subcylindrical, without distinctly dorsoventrally compressed lateral edges or serration; undivided terminal scansors on all fingers and toes, lamellae under 4th toe 8–9; preanofemoral scales 20–24. *Colouration* Dorsum cinnamon-brown, pale brown or greyish-brown; pair of longitudinal short, dark bars on neck, and 5–6 pairs of dark spots between axilla and groin along mid-dorsal region; tail usually reddish-brown, and often seen with tail curled. **HABITAT AND HABITS** Recorded from urban habitat in Ubud. Nocturnal and arboreal. Nothing more known of its biology, but it could be assumed to be similar to the more widespread **Mourning Gecko** *L. lugubris*.

Adult and hatchling (inset).

LACERTIDAE
WALL LIZARDS OR LACERTIDS

Lacertidae is a diverse family of lizards with nearly 340 species native to Europe, Africa and Asia.

They generally have distinct conical heads, slender and depressed bodies, long tails and well-developed limbs (especially hind limbs). The scales are large and systematically arranged on the head, relatively small and sometimes granular on the back, and rectangular on the ventral side. Osteoderms are present on the head scales. Tail autotomy is common within the members. Most species are sexually dimorphic, with the males and females having different colours and patterns.

Lacertids occupy a wide range of habitats including forests, scrublands, grasslands and deserts. Most European and African species are terrestrial, but most Southeast Asian species are arboreal. One species, the Sawtail Lizard *Holaspis guentheri*, is known to glide using its broad tail and flattened body as an aerofoil. Most species are oviparous, but a handful are viviparous. At least eight species are parthenogenetic.

A single species is known from Bali.

Asian Grass Lizard ■ *Takydromus sexlineatus* SVL 8cm, TL 30cm
(*Bahasa Indonesia* Kadal Rumput; *English* Six-striped Long-tailed Grass Lizard)

DESCRIPTION *Physique* Body small, slender and elongated; head sharply pointed; tail prehensile and 3–5 times the length of SVL. *Scales* Large and systematically arranged head scales with osteoderms, small pointy scales beneath the chin resembling a beard; 4–6 rows of large, plate-like scales on dorsum; side of body with single row of large plate-like scales; ventrals rectangular; 2 femoral pores on each thigh. *Colouration* Dorsum with coffee-brown vertebral stripe; yellow (in males) or cream (in females) dorsal stripes from behind the eye to tail; dorsolateral stripe black or dark brown; males with white spots on sides; venter light green, white or cream. **HABITAT AND HABITS** Inhabits open grasslands and marshes with tall grasses. Entirely diurnal. Arboreal and very agile. When approached, they first remain completely still, and if continued, flee very fast to safety. Regularly seen arm-waving, possibly to communicate with each other. Feed on insects and millipedes, and can jump into the air to catch flying insects. Oviparous, laying 2–3 eggs in a clutch.

Adult female.

SCINCIDAE
SKINKS

With more than 1,670 members, this is the largest tetrapod (four-limbed) reptile family in the world (though some taxonomists have suggested the division of this large family into several smaller families). Skinks are almost near global in distribution and occupy all continents (except Antarctica), as well as most isolated islets in the world. Some species easily adapt to human habitats, which assists their dispersal.

Skinks show extraordinary variation in body form and size, with several groups independently having reduced or lost limbs and digits. Generally, their bodies are elongated, and covered with glossy, imbricated scales. The head scales are systematically arranged as shields, often with osteoderms. Some species living in relatively dry habitats have fused eyelids with a clear window to prevent moisture loss.

Skinks inhabit fossorial, terrestrial, arboreal and even semiaquatic habitats. They are capable of tail autotomy. Most are diurnal and sun loving. Diets and feeding habits are highly variable: while a majority of the species feed on arthropods, some have specialized diets that include plant matter. Most tropical species are oviparous, but some temperate ones are ovoviviparous.

At least 11 skink species are known from Bali and Nusa Penida. Several other species, including those new to science, are likely to occur in the islands. A museum specimen mistakenly identified as an Emerald Tree Skink *Lamprolepis smaragdina* from the Banja area in northern Bali is actually of an Olive Tree Skink *Dasia olivacea*.

Symmetrical head shields and glossy, imbricated body scales are characteristics of skinks. Shown here is a Common Sun Skink.

Balinese Snake-eyed Skink ■ *Cryptoblepharus balinensis* SVL 5cm, TL 10cm
(*Bahasa Indonesia* Kadal Bali Bermata Ular)

DESCRIPTION *Physique* Body small, slender, depressed; head medium with depressed snout; eyes large; lower eyelid fused with rudimentary upper eyelid to form fixed spectacle that is surrounded by granular scales; limbs reduced; 19–24 subdigital lamellae under 4th toe. *Scales* No postnasal; frontal almost equal in size to prefrontals that are usually in broad contact; supralabials 7; infralabials 5–7, supraciliaries 5–7, but usually 6; body scales in 24–28 rows around midbody. *Colouration* Dorsum light brown or golden-brown with blackish central line from snout to base of tail, which is bifurcated on neck and head; each side of body has dark band commencing from eye and backwards, which broadens at body where it contains lighter speckles; limbs dark and highly speckled with pale spots; tail brownish or bronzy above, and darker on sides; venter bluish-white. **HABITAT AND HABITS** Known from dry western and northern parts of Bali. Arboreal and actively moves on tree trunks, though occasionally found on the ground. Diurnal, and feeds on insects and arachnids. Oviparous.

Adult colouration.

Beach Snake-eyed Skink ▪ *Cryptoblepharus cursor* SVL 4cm, TL 9cm
(*Bahasa Indonesia* Kadal Pantai Bermata Ular)

DESCRIPTION *Physique* Body small, slender, depressed; head medium with depressed snout; eyes large; lower eyelid fused with rudimentary upper eyelid to form fixed spectacle that is surrounded by granular scales; limbs relatively long; 18–19 subdigital lamellae under 4th toe. *Scales* Postnasal and supranasal absent; frontal almost equal in size to prefrontals that are usually in broad contact; supralabials 7; infralabials 5–6; supraciliaries 5; 3 enlarged supraciliaries; 24–26 rows around midbody. *Colouration* Dorsum with broad brown vertebral zone bordered by indistinct and narrow (sometimes also discontinued) black stripe, then more prominent and broader cream-coloured dorsolateral stripe; darker lateral sides specked with whitish spots; pale mid-lateral stripe extends from lips to back limbs; pattern faded on latter part of body; head and tail mostly bronzy-brown; venter silvery-white.
HABITAT AND HABITS Inhabits lowlands. Diurnal, but most active at dawn and dusk. Mostly shows terrestrial tendencies, and seen in littoral zone moving actively among debris on beaches. Oviparous, but reproductive habits unstudied.

Adult colouration.

Blue-tailed Snake-eyed Skink ■ *Cryptoblepharus renschi* SVL 4cm, TL 9cm
(*Bahasa Indonesia* Kadal Berekor Biru)

DESCRIPTION *Physique* Body small, slender, depressed; head medium with acute snout; eyes large; lower eyelid fused with rudimentary upper eyelid to form fixed spectacle that is surrounded by granular scales; limbs relatively short; 18–25 subdigital lamellae under 4th toe. *Scales* No postnasal (usually); frontal almost equal in size to prefrontals that are usually in broad contact; supralabials 7 (4 and 5 under eye); infralabials 6; supraciliaries 5–6; 3–4 enlarged supraciliaris; 2 nuchals; 20–28 rows around midbody. *Colouration* Dorsum blackish with 5 longitudinally aligned, bold, narrow, pale (whitish, cream-coloured or yellowish) stripes along body; vertebral stripe extends from snout to base of tail, top stripes on sides extend from above eye to tail, and bottom stripes from lips to back limbs; limbs dark and highly speckled with pale spots; tail olive or blue-grey, and more spotted than striped; venter bluish-white. **HABITAT AND HABITS** Known from Bali and Nusa Lembongan. Inhabits monsoon forests and savannah, as well as human habitats such as home gardens and walls in gardens. Largely arboreal, but (rarely) specimens are seen among leaf litter. Diurnal, and feeds on small insects and spiders.

Adult colouration and close up of head (inset).

Olive Tree Skink ■ *Dasia olivacea* SVL 11cm, TL 25cm
(*Bahasa Indonesia* Kadal Pohon Zaitun)

DESCRIPTION *Physique* Body robust; snout elongated and pointed; similar to *Eutropis* species in body form; eye opening small; tail slender and highly tapering. *Scales* Nuchals large; body scales slightly keeled and in 28–30 rows at midbody; males with 2 large heel scales; lamellae under 4th toe 17–22. *Colouration* Dorsum olive-brown, green-brown or yellow-olive; vague cross-bands comprising black-and-white dots may or may not be present on body and head (usually disappear with age); throat and venter light green or cream coloured, and unpatterned; juveniles orangey or golden-yellow, with bright and dark cross-bands, each wider than interspace; tail uniform bronze or orangey; limbs usually banded. **HABITAT AND HABITS** Inhabits open forests and plantations with large trees like coconut. Diurnal and highly arboreal. Feeds on insects and other arthropods. Oviparous, with clutch size of up to 14 eggs, usually laid among arboreal plants such as ferns, or in a tree-hole.

Adult and juvenile (inset).

Mangrove Skink ■ *Emoia atrocostata* SVL 9.5cm, TL 23cm
(*Bahasa Indonesia* Kadal Bakau; *English* Littoral Whiptail-skink)

DESCRIPTION *Physique* Body medium, slender, elongated; snout depressed, moderately long and tapering; eyes large; nostrils small; lower eyelid with clear window; limbs well developed; 32–42 subdigital lamellae under 4th toe. *Scales* Anterior loreal scale horizontally divided; prefrontals separated or narrowly in contact; interparietal narrow and distinct; frontoparietals fused; supralabials 6–8; infralabials 6–7; dorsals smooth. *Colouration* Dorsum greyish-olive to brownish; head often paler; body with dark and light flecks, sometimes merged to form bands or blotches; faint dark lateral stripe from face to tail; throat often bluish, sometimes with dark pigmentation or darker edges to the scales; venter whitish, yellowish or peach coloured. **HABITAT AND HABITS** Known from dry western parts of Bali. Inhabits mangroves, back-beach vegetation and rocky shorelines, seeking shelter in hollow tree trunks during high tide. Often ventures into the intertidal zone at low tide and has good swimming ability. Diurnal, and feeds on small crustaceans, fish, insects and other smaller lizards. Oviparous, with clutch size of 2 eggs, usually laid among driftwood or in tree-holes of mangrove trees.

Adult and juvenile (inset)

Common Sun Skink ■ *Eutropis multifasciata* SVL 12cm, TL 25cm
(*Bahasa Indonesia* Kadal Kebun, Kadal Tanah, Kadal Matahari)

DESCRIPTION *Physique* Body large and robust; head distinct with short snout; limbs well developed but relatively short; ear opening small with small and pointed lobules; tail less than twice length of body; female smaller in body size than male. *Scales* Lower eyelid scaly; paired enlarged nuchals; dorsals with 3 keels (larger scales may have 5 keels) and in 29–35

rows at midbody; lamellae under 4th toe 17–23. *Colouration* Dorsum bronze, brownish, olive-green or blackish; some individuals have 5–7 lines along back; sides darker with yellowish stripes, or series of white streaks or spots; males in breeding colouration have brighter red, orange or yellow sides; reddish-brown band on snout and black-and-white mottling on lower lip and sides of face; venter whitish; juveniles more brownish than adults, and prominently marked with stripes and/or spots on sides of body. **HABITAT AND HABITS** One of the most commonly seen reptiles in Bali and found in almost all habitat types, including within cities. Diurnal, terrestrial and sun loving. Sometimes basks communally. Feeds on insects, centipedes, other arthropods and occasionally small vertebrates. Ovoviviparous, giving birth to up to 10 young.

Juveniles (top) *and adults*.

Rough-scaled Sun Skink ■ *Eutropis rugifera* SVL 6cm, TL 12cm
(*Bahasa Indonesia* Kadal Matahari Bersisik Kasar)

DESCRIPTION *Physique* Body small and broad; head indistinct from neck and with short snout; limbs well developed but relatively short; ear opening small with small, pointed lobules; tail less than twice length of body. *Scales* Lower eyelid scaly; paired enlarged nuchals; dorsals with 3 or 5 prominent keels (larger scales may have 7 keels), and in 23–30 rows at midbody; lamellae under 4th toe 18–26. *Colouration* Dorsum olive-green, bronze, orange or brownish; some individuals have 5–7 yellowish longitudinal lines, or series of lighter spots, or black-edged pale mid-line along back; black-edged yellow stripe from eye backwards, broken into spots past shoulders in some individuals; male has bright red throat during breeding season; venter yellowish. **HABITAT AND HABITS** Usually found in edges of midland and lowland forests, mostly close to water. Diurnal, terrestrial and sun loving. Feeds on arthropods. Ovoviviparous.

Adult colourations.

Bowring's Supple Skink ▪ *Lygosoma bowringii* SVL 5cm, TL 12cm
(*Bahasa Indonesia* Kadal Bowring)

DESCRIPTION *Physique* Body slender and elongated; head triangular with tapering but blunt snout and nearly indistinct neck; ear opening rounded; limbs short, each with 5 fingers/toes; tail thick, cylindrical, tapering to narrow point. *Scales* Lower eyelid scaly and movable; single frontoparietal; paired nuchals; small supranasals present; paravertebral scales 52–58; dorsals smooth or slightly keeled, and in 24–28 rows on midbody. *Colouration* Dorsum bronze or light brown with fine darker longitudinal lines along body; dark line

from eye backwards; distinctive dark band on flanks with white dots; sides on front part of body have well separated or diffused white, yellow and pink (or reddish) zones, within which are black spots forming vague longitudinal lines; tail brown with pink or orangey sides in adults, bright red in juveniles; throat whitish; venter yellowish. **HABITAT AND HABITS** Mostly found in monsoon habitats, including agricultural land. Diurnal and semifossorial, living under leaf litter and loose soil. Feeds on small insects. Oviparous, with clutch size of 2–4 eggs.

Adult colourations.

Short-limbed Supple Skink ■ *Lygosoma quadrupes* SVL 8cm, TL 15cm
(*Bahasa Indonesia* Kadal Berkaki Pendek, Kadal Ular)

DESCRIPTION *Physique* Body extremely slender and elongated; head triangular, with tapering but blunt snout and nearly indistinct neck; ear opening rounded; limbs very short, each with 5 fingers/toes; tail thick (only slightly thinner than body), cylindrical, gradually tapering to narrow point. *Scales* Lower eyelid scaly and movable; supralabials 7; infralabials 6; no supranasals; single nasal; dorsals smooth and in 24–26 rows on midbody. *Colouration* Dorsum shades of brown or yellow-brown, with thin, dark longitudinal lines even on tail; forehead darker; venter and underside of tail pale with darker lines. **HABITAT AND HABITS** Mostly occurs in lowland forests, but also found in more open habitats such as plantations. Semifossorial, mainly living within and under decaying logs and humus-rich loose soil. Moves snake-like with limbs folded back to body when moving rapidly. Presumably feeds on termites and their eggs. Oviparous, with clutch size of 2–3 eggs.

Adult colourations.

Yellow-lined Forest Skink ■ *Sphenomorphus sanctus* SVL 5cm, TL 12cm
(*Bahasa Indonesia* Kadal Hutan Bergaris Kuning)

DESCRIPTION *Physique* Body small, slightly dorso-ventrally flattened; head triangular, with elongated but blunt snout and large eyes; limbs relatively long, each with 5 fingers/toes; tail thin and tapering to narrow point. *Scales* Supraoculars 5–7; supralabials 7 (5th and 6th under eye); infralabials 6; dorsals smooth or slightly keeled, and in 32–34 rows on midbody. *Colouration* Dorsum greyish-brown or blackish, with prominent yellow, gold

or whitish line from forehead to tail-tip; much broader bronze line on either side of body that merges with central yellow line at snout; lower flanks bronze, sometimes with white spots; limbs bronze or brownish and mottled; lips vaguely banded; venter whitish or yellowish. **HABITAT AND HABITS** Occurs in lowland rainforests and montane forests. Diurnal, but favours shade. Mostly arboreal, inhabiting large trees but also shows terrestrial activity, especially on rock faces and steep banks.

Adult colourations.

Van Heurn's Forest Skink ■ *Sphenomorphus vanheurni* SVL 6cm, TL 11cm
(*Bahasa Indonesia* Kadal Hutan Van Heurn)

DESCRIPTION *Physique* Body small, slightly dorso-ventrally flattened; head triangular, with elongated but blunt snout and large eyes; limbs relatively short, each with 5 fingers/toes; tail thick and tapering to narrow point. *Scales* Frontonasal twice as broad as long, forming broad, straight suture with rostral; loreals 2; supraoculars 4; supralabials 6 (4th under eye); dorsals smooth and in 31 rows on midbody; 4th toe covered above by 2 single scales at apex, followed by 4 paired scales, then lines of 3 scales each up to base of toe. *Colouration* Dorsum light brown with longitudinal lines made by series of dark brown dots; forehead mottled with dark brown; sides of head darker; upper lips with pale dots; flanks, sides of body and tail have white and black, minute spots or flecks; chin dark greyish with faint whitish mottling; venter yellowish; underside of tail whitish. **HABITAT AND HABITS** Occurs in montane forests in the lowlands and hills. Diurnal. Terrestrial or semifossorial, occupying places under debris on the ground or in leaf litter in shaded areas. Oviparous, and lays 2 eggs in each clutch. **NOTE** It is possible that future taxonomic reviews will place this species under the genus *Tytthoscincus*, and that the Balinese subspecies (*balicus*) will be elevated to full species level. Despite earlier records, more recent morphological and molecular studies indicate that **Temminck's Skink** *Tytthoscincus temmincki* does not occur in Bali.

Adult colouration and ventral aspects (inset).

DIBAMIDAE
WORM LIZARDS

This family with 25 members, has a unique distribution, with one species (*Anelytropsis papillosus*) occurring in Mexico, and the rest (*Dibamus* spp.) being restricted to Southeast Asia and western New Guinea. Note that the common name 'Worm Lizards' also refers to the Amphisbaenia, or legless members of the family Anguidae.

The small bodies of worm lizards are well adapted to burrowing into soil, with a rigidly fused skull, large, plate-like scales on the snout, no external ears and vestigial eyes each covered with a scale. Males have small, flap-like back limbs, but females are limbless. The males' flap-like limbs are used to grasp females during mating.

Worm lizards are semifossorial and live in loose soil, or under rocks or felled rotting logs on the forest floor. They are oviparous, and their eggs contain hard calcified shells. Tail autotomy is known in this group.

A single species is known from Nusa Penida, which also likely occurs in Bali.

A thick and large rostral, lack of external ears, and eyes each covered with a transparent scale, are characteristics of dibamids that live a semifossorial life. This is a close-up of the head of a Taylor's Oriental Worm Lizard.

Taylor's Oriental Worm Lizard ▪ *Dibamus taylori* SVL 10cm, TL 12cm
(*Bahasa Indonesia* Kadal Taylor)

DESCRIPTION *Physique* Body small, cylindrical and slender; head with blunt, round snout; eye reduced and covered by scale; no external ear opening visible; nostrils visible but tiny; no obvious limbs, but males have flap-like, vestigial back limbs on either side of cloaca. *Scales* Large rostral; small frontal; postoculars 3–4; body scales smooth and in 22–28 rows at midbody; ventrals more or less same size as dorsals; subcaudals 41–55, with males having longer tails than females. *Colouration* Dorsum usually unpatterned dark brown or purplish-brown; venter slightly paler; snout-tip, sides of head and throat may be lighter; some individuals have pale spots or mottling; tail-tip lighter. **HABITAT AND HABITS** Occurs in leaf litter-rich, tropical semi-deciduous forests and cropland. Fossorial, and presumably feeds on earthworms. Lays calcified eggs, but reproductive habits are not known.

Adult colouration.

VARANIDAE
MONITOR LIZARDS

Members of this group are found in the tropics and subtropics of the Old World, in Asia, Australia and Africa. Eighty members are included in the family.

The family includes the largest lizard species in the world. They have relatively small heads with long, slender necks and strong, muscular tails. Their limbs are well developed, with strong claws on each finger and toe. The body has juxtaposed scales on thick, loose-fitting skin. There are numerous juxtaposed, granular scales on the head. The tongue is slender, very long and deeply forked at the tip. Varanids are not capable of tail autotomy.

Monitor lizards inhabit a wide variety of habitats, and the family comprises terrestrial, arboreal and semiaquatic species. Most are actively foraging carnivores consuming large prey, but some arboreal species in the Philippines are seasonally frugivorous. They are oviparous and lay soft-shelled eggs.

One varanid species occurs in Bali.

The Asian Water Monitor is among the largest lizards in the world.

Asian Water Monitor ■ *Varanus salvator* SVL 70cm, TL 180cm
(*Bahasa Indonesia* Biawak Air; *Balinese* Aalluu)

DESCRIPTION *Physique* Body very large and robust in adults, especially males (it may attain more than 300cm in total length), and relatively slender in juveniles; head has depressed snout; rounded or oval nostrils, twice as far from orbit as from snout-tip; tail compressed, with sharp, double-toothed crest above. *Scales* Scales on forehead flat and smooth, and larger than strongly keeled nuchal scales. *Colouration* Dorsum drab greyish or blackish, with yellow, orange or white spots, or ocelli, in transverse series which are more prominent in juveniles than in adults; juveniles also have black bars on lips and yellow bands on tail. **HABITAT AND HABITS** Semiaquatic, occuring in swamps, ditches, streams, reservoirs, ponds, mangroves, paddy fields and (infrequently) near-coast shallow seas. Agile climber, and hatchlings spend most of their time in trees. Diurnal. Scavenger feeding on carcasses and carrion, but also actively predates on animals ranging from insects (by juveniles) and eggs of other animals, to medium-sized mammals and birds. Oviparous, laying up to 30 soft-shelled eggs in a burrow inside a termite mound, a tree-hole or a hole in a riverbank. **WARNING** Provoked animals may use their teeth, claws and tail to defend themselves, and all three can inflict significant injuries to humans.

Adult and juvenile (inset).

ACROCHORDIDAE
FILE SNAKES

This unique snake family consists of only three species, which are restricted to Asia and northern Australia.

File snakes have stout bodies covered with granular, rasp-like scales with projecting spines (thus the name 'file snakes'). The skin appears loose (hence the other common name, 'cloth snakes'). The tail is narrow but slightly laterally flattened.

This group is highly aquatic and inhabits both fresh water and coastal marine environments. The snakes are well adapted to swimming and crawling on muddy bottoms, but extremely sluggish on land. They predominantly feed on fish, and use their spiny skin to firmly grip on to slippery prey. They are mostly nocturnal but occasionally move around during the day. File snakes can remain submerged under water for long periods of time. They are ovoviviparous, with long gestation periods.

All species are non-venomous, but some of the bigger ones are capable of inflicting painful bites.

A single species of file snake is known from Bali.

Due to their banded colouration and the habitat they occupy, File Snakes are commonly misidentified as venomous sea snakes.

Little File Snake ■ *Acrochordus granulatus* TL ♀ 80m, ♂ 70m
(*Bahasa Indonesia* Ular Air Tawar, Ular Karung; *English* Wart Snake, Cloth Snake)

DESCRIPTION *Physique* Body stout and slightly compressed; head small with short snout and indistinct neck; female has relatively larger head than male; nostrils located dorsally and can be closed by cartilaginous flap when diving; eyes tiny and with vertical pupils; tail short and laterally compressed. *Scales* Body scales wart-like and with keels; about 100–150 scale rows at midbody, juxtaposed; ventrals reduced to median ridge; skin appears loose. *Colouration* Dorsum brown or grey with alternating pale buff bands, which are more distinct and zebra-like in juveniles than in adults; bands become faint and obscure as an animal grows, and older individuals appear uniformly deep brownish-black; sides can have tinge of reddish-brown; forehead sometimes has a few cream or pale-coloured spots.

HABITAT AND HABITS Inhabits brackish waters (mangroves, estuaries and similar) and coastal marine habitats, but occasional found in freshwater bodies. Largely nocturnal, but sometimes active during day. In daytime, lives in crab holes or remains burrowed in mud, occasionally coming out to breathe; or may stay anchored with snout above level of water. Physiologically adapted for long aerobic diving (even lasting up to 2 hours). Mostly crawls sluggishly over the bottom, but also swims. Feeds on bottom-dwelling gobies and goby-like fishes, though crustaceans, other snakes and even carrion are occasionally taken. Uses its rough skin to hold on to slippery prey. Ovoviviparous, giving birth to up to 12 young. **VENOM** Non-venomous and inoffensive snake. Commonly mistaken for true sea snakes due to colour pattern and habitat it lives in.

Adult colourations.

PYTHONIDAE
PYTHONS

With more than 40 members, this ancient Old World snake lineage consists of both very small (less than 1m in TL) and also the largest snakes in the world, some reportedly exceeding 6m in length. They naturally occupy equatorial or subequatorial regions of Africa, Asia and Australia. The family name derives from the word 'Python' the name of the giant snake killed by Apollo in Greek mythology.

All members of this family are non-venomous and kill prey by constriction. They use their sharp, backwards-curving teeth, four rows in the upper jaw, two in the lower, to grasp and restrain prey, then wrap a number of coils around it. Death mainly occurs by asphyxiation (severe deficient supply of oxygen to the body), and possibly also due to circulatory failure, but not by crushing to death as commonly believed. All prey is swallowed whole, and it may take a python several days or even months to fully digest it. A few members of this group are known predators of humans, though fatal attacks are very rare.

Pythons have infrared-sensitive organs called labial pits on their lip scales. These are extremely sensitive to small changes in temperature and are used to locate warm-blooded prey, and also possibly the approach of warm-blooded predators. Pythons possess anal spurs on each side of the cloaca, which are remnants of back limbs.

Pythons lay eggs (so are oviparous), and this differentiates them from boas (family Boidae), which give birth to live young (ovoviviparous). Females characteristically incubate the eggs by coiling themselves around them.

Two python species are known from Bali.

Labial pits on lip scales of a Reticulated Python.

Reticulated Python ■ *Malayopython reticulatus* TL 300–900cm
(*Bahasa Indonesia* Sanca Kembang, Ular Sanca Batik; *Balinese* Lélipi Saab, Lélipi Piton, Lélipi Ayé)

DESCRIPTION *Physique* Body thick, cylindrical and very large (this is the largest snake native to Asia); head lance shaped, containing rostral and labial pits, and with distinct neck; large eye with vertical pupil; short, blunt and prehensile tail; cloacal spurs more prominent in male than female.

Scales Dorsals smooth and in 69–79 rows at midbody; ventrals 297–330; subcaudals 75–102 and paired; anal undivided. *Colouration* Dorsum iridescent brown or yellowish with complex geometric pattern that incorporates different colours, including white, black, light brown and gold; head orange or yellowish with black median line; black line behind eye.

HABITAT AND HABITS Inhabits natural and man-made habitats in lowlands and midlands, mostly close to water. Displays terrestrial, arboreal and aquatic behaviours. Swims well, even in open ocean. Nocturnal. Feeds mainly on mammals ranging from rats to large prey such as pigs and even Sun Bears, but birds (especially fowl) and lizards are also occasionally taken. Frequently seen in caves looking for bats. Oviparous, with clutch size of up to 124 eggs. Facultative parthenogenesis is known in this species. Female incubates eggs by coiling herself around them.

VENOM Non-venomous snake, but one of the few snakes in the world known to eat humans occasionally. Given the large size of the snake, bites from it may cause severe lacerations.

Different colourations.

Different colourations, Reticulated Python.

Burmese Python ▪ *Python bivittatus* TL 200–300cm
(*Bahasa Indonesia* Sanca Bodo, Ular Sanca Burma)

DESCRIPTION *Physique* Body thick, cylindrical and very large; head lance shaped, containing rostral and labial pits, and with distinct neck; large eye with vertical pupil; short, blunt and prehensile tail; cloacal spurs more prominent in male than female. *Scales* Dorsals smooth and in 61–75 rows at midbody; ventrals 242–275; subcaudals 58–83 and paired; anal undivided. *Colouration* Dorsum light brown, yellowish-brown or greyish with large, irregularly shaped, golden-brown or dark brown, semi-quadrangular patches with dark edges in reticulated pattern along body; distinct lance-shaped mark extends from neck to middle of head; sides of face dark brownish with two lighter bands from eye to lips on each side. **HABITAT AND HABITS** Occurs at relatively low elevations, and found in forested as well as human habitats. Nocturnal, terrestrial snake, but also displays diurnal, arboreal and aquatic habits. Feeds on medium and large mammals (including livestock), birds and large reptiles such as monitor lizards. Possibly aestivates during dry season. When threatened, front part of body bunches into sinuous curves, and the snake sometimes hisses aloud, lashing out and biting savagely. Oviparous, with clutch size of up to 60 eggs laid inside a rock cave or large tree-hole. Female incubates eggs by coiling herself around them. Parthenogenetic reproduction is known in this species. **VENOM** Non-venomous snake. However, given its large size, bites from it may cause severe lacerations.

Adult colouration.

Adult colouration, Burmese Python.

XENOPELTIDAE
SUNBEAM SNAKES

This unique family of snakes consists of only two species. They are restricted to Southeast Asia, but are most closely related to neotropical sunbeam snakes of the family Loxocemidae, and also to pythons of the family Pythonidae.

Sunbeam snakes have subcylindrical bodies covered with highly iridescent, smooth scales that create a beautiful display of colours in the light. The forehead is covered with large scales and the snout is depressed. The tail is short.

The snakes are semifossorial and live under litter and loose soil in sheltered places. They predominantly feed on small vertebrates that are killed by constriction, and are mostly nocturnal. Both species are oviparous. Members of this family are non-venomous and quite timid.

A single species is known from Bali.

The highly iridescent, smooth scales of a Sunbeam Snake.

Sunbeam Snake ■ *Xenopeltis unicolor* TL 110cm
(*Bahasa Indonesia* Ular Pelangi; *Balinese* Lélipi Léngis, Lélipi Bebek, Lélipi Bianglalah)

DESCRIPTION *Physique* Body thick and cylindrical; head depressed, with wedge-shaped snout and slightly distinct neck; small eye with round pupil; short tail. *Scales* Dorsals smooth and in 15 rows at midbody; ventrals 164–196; subcaudals 24–31 and paired; anal divided. *Colouration* Dorsum iridescent and uniform brown, with lighter edges to the scales; juveniles have a pale collar; venter whitish. **HABITAT AND HABITS** Inhabits lowland and midland forests and human habitats, mostly near water. Semifossorial, occupying crevices, rodent burrows and places under leaf litter. Nocturnal and relatively slow moving. Feeds on amphibians, lizards, other snakes, birds and mammals (mainly rodents). Kills prey by constriction. May vibrate tail when threatened. Oviparous, with clutch size of 3–17 eggs. **VENOM** Non-venomous snake.

Adults and juvenile (bottom right).

COLUBRIDAE
COLUBRID SNAKES ('COMMON SNAKES')

With more than 1,940 members, this is the largest snake family in the world. They occupy all continents (except Antarctica), as well as several isolated islets.

Colubrids vary largely in body size and form, and are impossible to define using only external characteristics. They range from minute burrowing forms and slender arboreal species, to large-bodied terrestrial forms. In morphology they are similar to elapids, and both groups have large, symmetrically arranged head shields, broad ventral scales and fewer than 30 midbody scale rows. However, unlike elapids, most colubrids have a loreal scale between the nostril and the eye, and different dentition.

Colubrids occupy all habitats and feed on a variety of prey, ranging from snails and slugs to large mammals. Some species constrict their prey. A few live in groups at least seasonally (for example, they may be gregarious at times of mating, or hibernate in shared refugia). This group includes both oviparous and ovoviviparous species.

While most colubrids are non-venomous, some are mildly venomous, and a small number are highly venomous and have caused human fatalities. The venomous colubrids have an 'opisthoglyphous' fang assemblage, where a pair of enlarged teeth at the back of the maxillae, which normally angle backwards, have a groove to take venom into a puncture (thus the common name 'rear-fanged snakes'). The venom is produced in a primitive form of a venom gland called the 'Duvernoy gland'.

At least 24 species of colubrid are known from Bali and Nusa islands. Records of the Short-tailed Reed Snake *Calamaria virgulate* from the island need further verification.

The Mangrove Cat Snake is thought to have been introduced to Bali.

Green Vine Snake ▪ *Ahaetulla prasina* TL 130cm
(*Bahasa Indonesia* Ular Pucuk; *Balinese* Lélipi Busung, Lélipi Gadang Arjuné; *English* Oriental Whipsnake)

DESCRIPTION *Physique* Body long, slender and slightly compressed; elongated head with prominent groove along snout (termed 'canthus rostralis'); large eye with horizontal pupil; long, slender tail has prehensile tip. *Scales* Dorsals smooth and in 15 rows at midbody; ventrals 189–241; subcaudals 141–235 and paired; anal divided. *Colouration* Adult dorsum colouration varies from several shades of green (including fluorescent green) to brown, yellowish, bluish and even dark grey; body usually speckled with black spots, and with yellow or white stripe along edges of ventrals; interstitial skin of neck black and white.

HABITAT AND HABITS Arboreal and diurnal, and usually found on low vegetation, often near streams and other water sources. Sleeps on shrubs and at tips of branches in trees. Prehensile tail used to cling to branches while moving on vegetation. Lifts up large section of front body when moving on the ground. When provoked, coils up, raise itself, inflates front part of body, arches neck, opens mouth widely and may strike. Feeds on amphibians, lizards and birds, and hunts by ambush using its unusual binocular vision. Ovoviviparous, giving birth to up to 12 young. **VENOM** Rear-fanged (ophisthoglyphous) snake. Bites may cause localized mild swelling and pain, which subsides very fast. Rarely, may take a few days to do so.

Different colourations.

Different colourations.

Dog-toothed Cat Snake ■ *Boiga cynodon* TL 200cm

(*Bahasa Indonesia* Ular Blidah, Ular Kucing Bergigi Panjang, Ular Bajing; *Balinese* Lélipi Sélang Ambu, Lélipi Awan)

DESCRIPTION *Physique* Body long, slender and slightly laterally compressed; head subovate with short, round snout and distinct neck; teeth in front and lower jaw noticeably long (thus the common name); eyes large with vertical pupils. *Scales* Dorsals smooth, in 23 or 25 rows at midbody, and those in vertebral row are strongly enlarged; ventrals 248–290; subcaudals 114–165 and paired; anal undivided. *Colouration* Dorsum yellowish-brown, greyish-brown or brownish-tan, with dark brown or reddish-brown cross-bands; some individuals have whitish bands in between dark ones; head usually lighter and yellowish (especially lip scales) than rest of body, and with fine, darker stripe extending from behind eye to neck or jaw; melanistic individuals are not uncommon.

HABITAT AND HABITS Mostly arboreal and occurs in wide range of habitats, including primary and secondary forests, cultivated places, and even home gardens and urban areas. Found on lower vegetation as well as in the canopy. Sometimes congregates in cave systems. Nocturnal. Feeds mostly on birds and birds' eggs, but may also take amphibians, reptiles and mammals. Envenomation and constriction used in prey capture, and tail can also be used to handle prey. Oviparous, with clutch size of up to 23 eggs. **VENOM** Rear-fanged (ophisthoglyphous) snake. Localized swelling and pain from bites of smaller individuals usually subsides very fast, but bite effects from larger specimens may last for hours or even days. **NOTE** The **Lesser Sundas Cat Snake** *B. hoeseli*, known from the Nusa Tenggara Islands, is similar in colouration and may occur in Bali.

Adult colouration.

Juvenile (top) *and adult* (bottom).

Mangrove Cat Snake ▪ *Boiga dendrophila* TL 250cm
(*Bahasa Indonesia* Taliwangsa, Ular Cincin Emas, Cincin Perak; *Balinese* Lélipi Sabuk, Lélipi Bungkung)

DESCRIPTION *Physique* Body robust and stout, and laterally compressed; head subovate with short, round snout and distinct neck; eyes large with vertical pupils. *Scales* Dorsals smooth, in 21 or 23 rows at midbody, with those in vertebral row strongly enlarged; ventrals 209–253; subcaudals 89–118 and paired; anal undivided. *Colouration* Dorsum glossy black with 21–45 narrow yellow or whitish cross-bands along body and about 10 on tail; bands usually do not meet over back. Throat and lower jaws yellowish with black edges in scales.

HABITAT AND HABITS Presumed to be have introduced to Bali recently from other

parts of Indonesia. Mostly arboreal, and inhabits vegetated habitats close to water (riverine forests, paddy fields, mangroves, peat swamps and similar). Hunts both on vegetation and on land. Nocturnal. Feeds mostly on birds and birds' eggs, but may also take amphibians, lizards, other snakes and even medium-sized mammals such as mouse deer. Oviparous, with clutch size of 4–15 eggs. **VENOM** Rear-fanged (ophisthoglyphous) snake. Mild local swelling, pain (and rarely slight fever) after bite usually subside in a few minutes, but rarely can last for a few days. The venom has been studied well and is known to contain several bird-specific toxins.

Adult colourations.

■ Colubrid Snakes ('Common Snakes') ■

Marbled Cat Snake ■ *Boiga multomaculata* TL 100cm
(*Bahasa Indonesia* Ular Kucing Loreng; *English* Many-spotted Cat Snake)

DESCRIPTION *Physique* Body long, slender and slightly laterally compressed; head subovate with short, round snout and distinctly slender neck; eyes large with vertical pupils. *Scales* Dorsals smooth and in 17 or 19 rows at midbody; ventrals 196–245; subcaudals 72–111 and paired; anal undivided. *Colouration* Dorsum greyish-brown with double series of brown spots (often edged with thicker black and thinner white margins) alternating along back; smaller spots on flanks; venter greyish and spotted with brown; brown line from snout to jaws across eye and back line behind eye; 'V'-shaped mark on head. **HABITAT AND HABITS** Mostly arboreal and occurs in lowland forests, open bushland and bamboo thickets, as well as home gardens. Usually encountered close to water. Nocturnal. Mainly feeds on lizards, but also takes other vertebrates. Forages both on vegetation and on the ground, and feeds on lizards and birds. Oviparous, with clutch size of 4–8 eggs. When threatened, opens mouth widely and inflates and coils front part of body into loops before striking. **VENOM** Rear-fanged (ophisthoglyphous) snake. There are no records of medically significant symptoms after bites from this species.

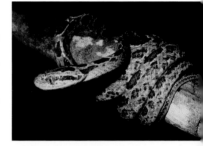

Adult colourations.

Black-headed Cat Snake ■ *Boiga nigriceps* TL 120cm
(*Bahasa Indonesia* Ular Kucing Hitam Kepala)

DESCRIPTION *Physique* Body long, slender and slightly laterally compressed; head subovate with short, round snout and distinct neck; eyes large with vertical pupils. *Scales* Dorsals smooth and in 21 rows at midbody; ventrals 240–293; subcaudals 134–164 and paired; anal undivided. *Colouration* Dorsum ranges from greyish to reddish-brown to bright red; head and neck darker and mostly olive or brown; throat and lower lips yellowish or cream; some individuals have small black dots on flanks. **HABITAT AND HABITS** A

record exists from north of Balian area. Arboreal, and inhabits lowland forests. Nocturnal. Forages mostly on vegetation and feeds on lizards, other snakes, birds and mammals. Oviparous, with clutch size of up to 3 eggs. **VENOM** Rear-fanged (ophisthoglyphous) snake. Bite causes mild local symptoms, including minor swelling and pain.

Adult colourations.

Cuvier's Reed Snake ■ *Calamaria schlegeli* TL 40cm

(*Bahasa Indonesia* Ular Cuvier, Ular Lemas, Ular Peliang; *English* Red-headed Reed Snake)

DESCRIPTION *Physique* Body cylindrical and slender; head short with indistinct neck; eyes small and with round pupils; short and blunt tail. *Scales* Dorsals smooth and in 13 rows at midbody; ventrals 130–180; subcaudals 19–44 and paired; anal undivided. *Colouration* Dorsum black or dark brown, and venter yellowish; forehead dark brown in subspecies *C. s. cuvieri* found in Bali and Java (red or orange in subspecies *C. s. schlegeli* occuring on other islands and Southeast Asian mainland). **HABITAT AND HABITS** Terrestrial and semifossorial. Often found among leaf litter in lowland forests and plantations. Nocturnal. Feeds on frogs, arthropods, earthworms and slugs. Oviparous. **VENOM** Non-venomous snake.

Adult colouration.

Paradise Tree Snake ■ *Chrysopelea paradisi* TL 100cm
(*Bahasa Indonesia* Ular Pohon Paradisi, Ular Terbang; *English* Garden Flying Snake)

DESCRIPTION *Physique* Body long and slender; head depressed, pear shaped and with distinct neck; large eyes with round pupils; long and slender tail has prehensile tip; skin has increased strength and energy storage associated with higher strain rates, enabling stretching during gliding. *Scales* Dorsals smooth or slightly keeled, with apical pits at free ends, and in 17 rows around midbody; ventrals 198–239, with distinct keels laterally; subcaudals 106–149 and paired; anal divided. *Colouration* Adult dorsum black with green spot in each scale; some adults have orange, pink or red spots along top of body; forehead has yellow bands; lips yellowish; juveniles brightly coloured. **HABITAT AND HABITS** Arboreal and diurnal. Snakes of the *Chrysopelea* genus are the only limbless animals that glide through air using their flattened bodies. It is likely that gap-crossing between trees may have been the evolutionary reason for gliding in these snakes. Feeds mostly on lizards, but also on birds and small bats. Constricts prey. Oviparous, with clutch size of up to 8 eggs. **VENOM** Rear-fanged (ophisthoglyphous) snake. No records of medically significant symptoms after bites.

Adult (left) and juvenile (right).

Yellow-striped Racer ■ *Coelognathus flavolineatus* TL 180cm

(*Bahasa Indonesia* Ular Babi, Ular Kopi, Ular Racer Berbelang Kuning; *Balinese* Lélipi Kopi; *English* Yellow-striped Trinket Snake)

DESCRIPTION *Physique* Body long and slender; head elongated, with long snout and slightly distinct neck; large eyes with round pupils; long and slender tail. *Scales* Dorsals slightly keeled and in 19 rows around midbody; ventrals 193–242; subcaudals 80–116 and paired; anal undivided. *Colouration* Adults metallic brown on first half of dorsum, grading to dark blue-grey on second half; faint blotches or streaks across body; dark stripe from back of eye to end of jaw, and another on nape of each side; juveniles have bright yellow mid-dorsal stripe edged with black running almost half length of body, and cross-bands consisting of alternating white and black spots. **HABITAT AND HABITS** Lives in forested habitats, as well as agricultural systems and home gardens. Displays both terrestrial and arboreal behaviours. Feeds on amphibians, lizards and mammals (mainly rodents). Inflates throat and coils body into 'S'-shaped loops when threatened. Oviparous, with clutch size of up to 12 eggs. **VENOM** Non-venomous snake.

Adult and close-up of head (inset).

Copperhead Racer ■ *Coelognathus radiatus* TL 170cm
(*Bahasa Indonesia* Lanang Sapi, Sapi Lanang, Ular Tikus, Ular Racer Berkepala Tembaga;
Balinese Lélipi bikul; *English* Copperhead Trinket Snake)

DESCRIPTION *Physique* Body long and slender; head elongated, with long snout and slightly distinct neck; large eyes with round pupils; long and slender tail. *Scales* Dorsals smooth in front part of body and flanks along body, weekly keeled in dorsal front part of body, and in 19 rows around midbody; ventrals 207–250; subcaudals 80–108 and paired; anal undivided. *Colouration* Adult dorsum greyish or yellowish-brown with 2 broad and 2 narrow longitudinal black stripes on front part of body; some individuals have whitish cross-bands on front part of body; head coppery-brown with 3 black lines radiating down and backwards; juveniles more yellowish than adults, and usually without white cross-bands. **HABITAT AND HABITS** Occurs in open forests, grassland, agricultural fields, and suburban and urban areas. Displays both terrestrial and arboreal behaviours. Feeds on amphibians, lizards, birds and mammals (mainly rodents). Inflates throat, opens mouth widely, coils front part of body into 'S'-shaped loops and raises itself when threatened. Oviparous, with clutch size of 5–23 eggs. **VENOM** Non-venomous snake.

Subadult.

Adults in threat display.

Lesser Sundas Bronzeback ■ *Dendrelaphis inornatus* TL 100cm
(*Bahasa Indonesia* Ular Tali)

DESCRIPTION *Physique* Body long and slender; head elongated and pear shaped, with distinct neck; large eyes with round pupils; long and slender tail has prehensile tip. *Scales* Dorsals smooth and in 15 rows around midbody; vertebrals enlarged; ventrals 182–208, with keel on either side that forms continuous ridge along bottom sides of body; subcaudals 122–163 and paired; anal divided (rarely entire). *Colouration* Dorsum dull brownish-green to olive-green; interstitial skin bright blue and visible when animal is excited and inflates the throat and body; narrow black stripe separates darker dorsal part of head from whitish or light yellowish lip scales; last lateral scale row lighter in colour, making a light longitudinal stripe along body; tongue dull brownish-red; venter yellowish-green or yellowish-white. **HABITAT AND HABITS** Known from Nusa Penida. Arboreal and diurnal, and often encountered on low vegetation in savannah and monsoon forests. Moves rapidly on vegetation. Lifts up large section of front body when moving on the ground. Feeds on lizards and frogs. Oviparous, with clutch size of up to 18 eggs. **VENOM** Non-venomous snake.

Adult and close-up of head (inset).

COLUBRID SNAKES ('COMMON SNAKES')

Painted Bronzeback ■ *Dendrelaphis pictus* TL 100cm
(*Bahasa Indonesia* Ular Tali, Ular Tampar Jawa, Ular Tali Picis, Lidah Api; *Balinese* Lélipi Angasan, Lélipi Jali)

DESCRIPTION *Physique* Body long and slender; head elongated, pear shaped and with distinct neck; large eyes with round pupils; long and slender tail has prehensile tip. *Scales* Dorsals smooth and in 15 rows around midbody; vertebrals enlarged; ventrals 163–208, with keel on either side that forms continuous ridge along bottom sides of body; subcaudals 99–169 and paired; anal divided. *Colouration* Dorsum brownish-olive or brownish-bronze; interstitial skin bluish, greenish or white, and visible when animal is excited and inflates throat and body; black-edged cream or yellow line runs along body at flanks on each side; black stripe from behind eye to neck; lips and chin yellowish or yellow-whitish; tongue red with blackish or darker tips. **HABITAT AND HABITS** Arboreal and diurnal, and often encountered on low vegetation and close to water. Moves rapidly on vegetation. Lifts up large section of front body when moving on the ground. Sleeps on shrubs and at tips of branches in trees. Feeds on lizards and frogs, and also hunts on the ground. Cannibalism has been noted in species. Oviparous, with clutch size of up to 8 eggs; multiple clutches a year. Known to lay eggs in tree holes as well as on the ground. **VENOM** Non-venomous snake.

Adult and close-up of head (inset).

Orange-bellied Snake ■ *Gongylosoma baliodeirum* TL 40cm
(*Bahasa Indonesia* Ular Tanah Bertotol-totol; *English* Spotted Ground Snake)

DESCRIPTION *Physique* Body short and slender; head short with blunt, round snout and slightly distinct neck; large eyes with round pupils. *Scales* Dorsals smooth and in 13 rows around midbody; ventrals 115–165; subcaudals 42–79 and paired; anal divided. *Colouration* Dorsum reddish-brown or dark brown with paired rows of well-spaced cream spots, which fade towards front part of body; lip scales have dark edges; some individuals have a few dark-edged white spots on flanks immediately behind neck; venter cream-orangey, sometimes with darker spots. **HABITAT AND HABITS** Occurs in lowland and submontane forests. Mostly terrestrial, spending time under leaf litter, but may (rarely) climb low shrubs. Usually nocturnal and crepuscular, but occasionally active during the day. Feeds on invertebrates (mostly spiders and insects), but also takes lizards. Oviparous, with clutch size of 2–3 eggs. **VENOM** Non-venomous snake.

Adult and close-up of head (inset).

Red-tailed Racer ▪ *Gonyosoma oxycephalum* TL 170cm
(*Bahasa Indonesia* Ular Racer Berkepala Tembaga, Ular Gadung Luwuk, Boncleng, Ular
Bakau Hijau; *Balinese* Lélipi bikul; *English* Copperhead Trinket Snake)

DESCRIPTION *Physique* Body long, slender and compressed; head elongated with long,
squarish snout, and distinct neck; large eyes with round pupils. *Scales* Dorsals smooth
or slightly keeled on back, and in 23, 25 or 27 rows around midbody; ventrals 203–263;
subcaudals 120–157 and paired; anal divided. *Colouration* Dorsum bright or light green
(occasionally emerald-green, yellowish, bluish or brownish); body scales have darker edges;
interstitial skin has dark and light areas; head more olive-yellowish, and with blackish
band from nostril to jaw across eye; tongue blue; lips yellowish or light green; venter
lighter green or yellowish; juveniles more olive than adults, with whitish cross-bands; tail
reddish or greyish-brown in adults, but more orangey or reddish in juveniles. **HABITAT
AND HABITS** Occurs in mangrove swamps, lower montane forests, agricultural fields, and
suburban and urban areas. Mostly arboreal, but also shows terrestrial behaviours. Diurnal.
Adult feeds on birds and mammals (mainly rodents but also bats) while juveniles take
lizards. Inflates throat, opens mouth widely, coils front part of body into 'S'-shaped loops
and rises when threatened. Oviparous, with clutch size of up to 12 eggs. **VENOM** Non-
venomous snake, though large individuals may inflict painful bites.

Adult and subadult (inset).

Common Wolf Snake ■ *Lycodon capucinus* TL 75cm
(*Bahasa Indonesia* Ular Cecak, Ular Genteng, Ular Rumah; *Balinese* Lélipi Cécék, Lélipi Jumahan; *English* Island Wolf Snake)

DESCRIPTION *Physique* Body small, slender and subcylindrical; head pear shaped and flat, with distinct neck; eye small with vertical pupil; tail short, round and tapering. *Scales* Preocular 1; dorsals smooth and in 17 (rarely 19) rows around midbody; ventrals 176–224; subcaudals 53–80 and paired; anal undivided. *Colouration* Highly variable within its larger range. In common form, dorsum is grey-brown, dark brown or purple-brown with light-coloured free edges; light edges form reticulated patterns or indistinct cross-bars along body; yellow, whitish or cream band across neck that may have brown spots within; lips whitish with brown spots; venter uniform white or cream. **HABITAT AND HABITS** Terrestrial and semiarboreal. Occurs in natural habitats and human habitations alike. Nocturnal, and rests under heaps of rubble, logs and leaf litter, inside crevices, underneath bark and in storerooms during day. When disturbed, hides head underneath coils of body, and sometimes vibrates tail. Also known to employ death feigning – in which an individual may turn almost half its body upside down, exposing its ventral scales, and remain in this position. Feeds voraciously on geckos, but also takes other lizards. Front teeth large and specialized for durophagy. Oviparous, with clutch size of up to 11 eggs. **VENOM** Non-venomous snake. However, most individuals bite savagely when handled.

Adult and hatchling (inset).

White-banded Wolf Snake ■ *Lycodon subcinctus* TL 100cm
(*Bahasa Indonesia* Ular Cecak Belang; *Balinese* Lélipi Cécék Poleng, Lélipi Tanah)

DESCRIPTION *Physique* Body small, slender and subcylindrical; head pear shaped and flat, with distinct neck; eyes small with vertically oval pupils; tail short, round and tapering. *Scales* No preoculars or rarely 1; dorsals slightly keeled (especially on vertebral row) and in 17 rows around midbody; ventrals 192–230; subcaudals 60–91 and paired; anal divided (rarely entire). *Colouration* Dorsum dark brown or black with up to 20 wide, white or cream bands across body; first band at neck and may cover gular area and lips; pattern most distinct in juveniles, and some older adults are completely bandless; venter white. Mimics the Malayan Krait (see p. 134) in colouration (though kraits have smaller heads and large, hexagonal scales along vertebrae).

HABITAT AND HABITS Terrestrial and arboreal. Occupies natural habitats and human habitations alike. Nocturnal, and rests under heaps of rubble, logs and leaf litter, inside crevices, underneath bark and within storerooms during day. When disturbed, hides head underneath coils of body, and sometimes vibrates tail. Feeds on geckos (hence commonly found around houses) and skinks, as well as frogs. Teeth specialized for durophagy. Oviparous, with clutch size of up to 11 eggs.
VENOM Non-venomous snake. However, most individuals bite savagely when handled.

Adult colourations.

Boie's Kukri Snake ■ *Oligodon bitorquatus* TL 45cm
(*Bahasa Indonesia* Ular Kukri Boie)

DESCRIPTION *Physique* Body very small; head ovate and not distinct from neck; eyes moderate with round pupils; tail short, round, tapering and ends in harder spine. *Scales* One long loreal present or absent; preocular 1; postoculars 2; dorsals smooth and in 17 rows around midbody; ventrals 130–165; subcaudals 28–46 and paired; anal undivided. *Colouration* Dorsal dark brown or dark purple, with yellow, orange or red spots arranged more or less as cross-bands; spots on vertical line are larger; head with alternate darker and lighter bands, often with darker band across eyes and lighter band on neck; venter reddish with blackish markings, sometimes as cross-bands. **HABITAT AND HABITS** Nocturnal and terrestrial. Occupies both natural and human habitats, especially farmland. No information on feeding and food available. Oviparous, with clutch size of 3–4 eggs. When disturbed, displays bright red ventral surface and tries to prick with tail spine. **VENOM** Non-venomous snake.

Adult in threat display and ventral aspects (inset).

Eight-striped Kukri Snake ■ *Oligodon octolineatus* TL 50cm
(*Bahasa Indonesia* Ular Kukri Bergaris Delapan, Ular Birang)

DESCRIPTION *Physique* Body small, slender and cylindrical; head ovate and not distinct from neck; teeth strongly curved backwards, in shape of a kukri knife used by Ghurka soldiers; eyes moderate with round pupils; tail short, round and tapering. *Scales* Dorsals smooth and in 17 rows around midbody; ventrals 155–197; subcaudals 43–61 and paired; anal undivided. *Colouration* Dorsum orange-brown or light brown with reddish vertebral stripe; 6–8 black longitudinal stripes along body; head with two distinct darker cross-bands, one across eyes. **HABITAT AND HABITS** Diurnal, but may also show strong crepuscular habits. Largely terrestrial, but also climbs. Inhabits lowland forests, as well as fringe areas of scrub and forests close to human habitats. Feeds on frogs, lizards, other snakes and eggs of frogs, reptiles and birds. Small eggs swallowed whole, while bigger ones slit using blade-like, curved teeth to help collapse and ingestion of egg contents. Oviparous, with clutch size of up to 5 eggs. **VENOM** Non-venomous snake.

Adult in threat display and close-up of head (inset).

Indo-Chinese Rat Snake ■ *Ptyas korros* TL 110cm
(*Bahasa Indonesia* Ular Sawa, Ular korros; *Balinese* Lélipi Sélém Réngas; *English* Javan Rat Snake)

DESCRIPTION *Physique* Body slender and very long (can grow up to about 250cm); head elongated, with distinct neck; eyes large with round pupils. *Scales* Supralabials 7–8; dorsals smooth on front part of body, slightly keeled on back of body, and in 15 rows at midbody; ventrals 160–187; subcaudals 120–151 and paired; anal divided. *Colouration* Dorsum greyish, brownish or occasionally reddish on front part of body, and darkening to near black on back of body; scales have whitish edges; hatchlings brownish with light grey cross-bands, made of spots and most prominent on front part of body. **HABITAT AND HABITS** Mostly lives in open habitats including forest edges, agricultural systems and home gardens.

Displays both terrestrial and arboreal behaviours. Feeds on amphibians, lizards, birds and their eggs, and mammals. Readily takes rodents, hence plays important role in pest control, especially in paddy industry. When provoked, flattens front part of body, expands throat region, and produces deep, rumbling defensive growl and hiss like that of a cobra. Oviparous, with clutch size of up to 14 eggs. **VENOM** Non-venomous snake. However, given its large size and aggressive nature, it may inflict painful bites.

Subadult colourations.

Banded Rat Snake ■ *Ptyas mucosa* TL 200cm
(*Bahasa Indonesia* Ular Jali, Ular Jali Belang, Ular Kayu; *Balinese* Lélipi Tiyih; *English* Oriental Rat Snake)

DESCRIPTION *Physique* Body slender and very long (can grow up to about 350cm); head elongated, with distinct neck; eyes large with round pupils. *Scales* Supralabials 8–9; dorsals slightly keeled or smooth, and in 17 rows at midbody; ventrals 187–213; subcaudals 95–146 and paired; anal divided. *Colouration* Dorsum colour varies largely from yellowish, reddish, greyish, brownish or blackish; dark bands or reticulated patterns usually present on latter part of body; lip scales have darker edges; hatchlings brownish with light grey cross-bands, most prominently on front part of body. **HABITAT AND HABITS** Mostly occupies open habitats, including agricultural systems and home gardens. Combats between males often seen, where each animal attempts to get its head and neck over the opponent and force it down. Catholic in diet, feeding on amphibians, birds and their eggs, mammals, other snakes, lizards and even turtles. Frequently forages in caves looking for bats. Readily takes rodents, hence plays important role in pest control, especially in paddy industry. When provoked, flattens front part of body, expands throat region and produces deep, rumbling defensive growl and hiss like that of a cobra. Oviparous, with clutch size of 5–25 eggs. Eggs usually laid in termite mound or tree-hole, and some females remain coiled around eggs during incubation. **VENOM** Non-venomous snake. However, given its large size and aggressive nature, it may inflict painful bites.

Adult and hatchling (inset).

Spotted Keelback ■ *Rhabdophis chrysargos* TL 70cm
(*Bahasa Indonesia* Ular Air Leher Merah, Ular Sapi, Ular Picung; *Balinese* Lélipi Sampi;
English Speckle-bellied Keelback)

DESCRIPTION *Physique* Body slender and moderate; head ovate with distinct neck; eyes large with round pupils. *Scales* Dorsals keeled and in 19 rows at midbody; ventrals 139–184; subcaudals 56–101 and paired; anal divided. *Colouration* Highly variable. Dorsum typically greyish-brown to olive-brown, with numerous equally spaced, dark, narrow cross-bands; bars dark on top and pale on flanks, and may contain lighter oblong marks within; head varies from orange-brown to grey to black; whitish lips with dark edges; backwards-pointing pale chevron on back of neck edged with black, which is more prominent in juveniles; yellow, orange-brown or reddish colouration behind neck in some

individuals. **HABITAT AND HABITS** Occurs in midland and upland wetland habitats, including paddy fields and other plantations close to water. Nocturnal and diurnal. Terrestrial, and mainly feeds on frogs and tadpoles, but lizards, fish, small birds and mammals are also taken. Oviparous, with clutch size of up to 10 eggs. **VENOM** Rear-fanged (ophisthoglyphous) snake. No records of medically significant symptoms after bites. However, several members of this genus (including the **Tiger Keelback** *R. tigrinus*) have caused human deaths.

Adult (top) and juvenile (bottom).

Striped Litter Snake ■ *Sibynophis geminatus* TL 50cm
(*Bahasa Indonesia* Ular Serasah; *English* Boie's Many-tooth Snake)

DESCRIPTION *Physique* Body slender and cylindrical; head short, flattened and with slightly distinct neck; eyes large with round pupils. *Scales* Dorsals smooth and in 17 rows at midbody, with tubercules but without pits; ventrals 140–183; subcaudals 73–145 and paired; anal divided. *Colouration* Highly variable. Dorsum dark brown with two yellowish or orange longitudinal lines along body; some individuals have bright orange collar on neck; yellowish lip scales with black sutures and sometimes dark mottling; venter greenish-yellow on front part of body and light green on latter part; black spots on edges of ventrals form narrow stripe along lower flanks. **HABITAT AND HABITS** Occurs in forest and open habitats in lowlands and hills, mostly close to water. Diurnal and terrestrial. Primarily feeds on skinks, but probably also takes amphibians and invertebrates. Oviparous, with clutch size of up to 3 eggs. **VENOM** Non-venomous snake.

Adult colourations.

Javan Keelback ■ *Xenochrophis melanozostus* TL 80cm
(*Bahasa Indonesia* Ular Bandotan Tutul; *Balinese* Lélipi Yeh, Lélipi Amin)

DESCRIPTION *Physique* Body robust and cylindrical; head slightly oval and with distinct neck; eyes large with round pupils. *Scales* Dorsals keeled and in 19 rows at midbody; ventrals 128–142; subcaudals 66–83 and paired; anal divided. *Colouration* Dorsum brown, with 'blotched form' having elongated blotches, and 'striped form' having broad, dark, longitudinal stripes; widely open 'V'- or 'U'-shaped mark on neck, and 2 well-defined subocular streaks on head; ventrals and subcaudals have dark margins. **HABITAT AND HABITS** Semiaquatic snake found in wide variety of wetland habitats, including paddy fields. Predominantly diurnal, but nocturnal activity has also been noted. Sometimes sleeps on low vegetation. Feeds chiefly on frogs (especially ranids) and fish. When threatened, flattens neck and front part of body (possibly imitating a cobra) and may bite savagely. Also displays caudal autotomy and death feigning. Oviparous. **VENOM** Non-venomous snake.

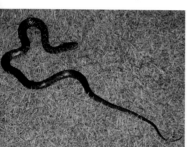

Adult colourations.

Red-sided Keelback ■ *Xenochrophis trianguligerus* TL 120cm
(*Bahasa Indonesia* Macan Air Segitiga; *English* Triangle Keelback)

DESCRIPTION *Physique* Body slender and cylindrical; head large, slightly oval and with distinct neck; eyes large with round pupils. *Scales* Single preocular; postoculars 3; supralabials 9; infralabials 10; dorsals keeled except for the outermost rows and in 19 rows at midbody; ventrals 132–150; subcaudals 86–105 and paired; anal divided. *Colouration* Dorsum olive-brown or blackish-brown, with regularly spaced orange or reddish triangle marks on sides of neck and anterior part of the body; markings are more brownish or greyish in older animals; dorsum with triangle-shaped dark markings; head olive above; lip scales yellow and edged with black; ventral yellowish or cream. **HABITAT AND HABITS** Semiaquatic snake found in wide variety of wetland habitats, including paddy fields. Predominantly nocturnal. Feeds chiefly on frogs (especially ranids) but also rodents and small birds. Oviparous, with clutch size of 5–15 eggs. **VENOM** Non-venomous snake.

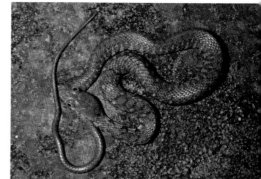

Subadult colourations.

Striped Keelback ■ *Xenochrophis vittatus* TL 60cm
(*Bahasa Indonesia* Ular Lare Angon; *Balinese* Lélipi Carik)

DESCRIPTION *Physique* Body slender and cylindrical; head slightly oval and with distinct neck; eyes large with round pupils. *Scales* Dorsals mostly keeled and in 19 rows

at midbody; ventrals 140–151; subcaudals 53–84 and paired; anal divided. *Colouration* Dorsum brown with 3 thick, longitudinal black lines along body, and sometimes with thinner black lines in between thicker ones; head light in colour and heavy mottled with dark marks; lip scales whitish with dark edges; ventral edges dark in colour. **HABITAT AND HABITS** Semiaquatic snake found in wide variety of wetland habitats, including paddy fields. Predominantly diurnal, but nocturnal activity has also been noted. Feeds chiefly on fish. Oviparous, with clutch size of 3–11 eggs. **VENOM** Non-venomous snake.

Adult colourations.

LAMPROPHIIDAE
SAND SNAKES, HOUSE SNAKES AND SIMILAR

This is a loosely bound family of more than 325 species, primarily found in Africa and with a few species extending into Europe, America, the Middle East and Asia. Some members of the family were once placed under the Colubridae.

Lamprophiids are an ecologically diverse group. They include both diurnal and nocturnal species, as well as fossorial, terrestrial and aquatic snakes. Some fast-moving species actively hunt their prey, while others are sit-and-wait predators. Some lack a loreal, and have smooth, shiny scales.

Lamprophiids are non-venomous snakes and are not considered dangerous to humans

Two lamprophiid species are known from Bali.

The lamprophiid Common Mock Viper mimics viperids in behaviour and morphology.

Common Mock Viper ■ *Psammodynastes pulverulentus* TL 60cm
(*Bahasa Indonesia* Ular Viper Tiruan, Ular Beludak Palsu)

DESCRIPTION *Physique* Body short and slender; head flattened and triangular, with truncate and short snout, and distinct neck; snout-tip of adult slightly curved up; eyes large; pupil round or slightly elliptical when unthreatened, but constricts and shifts to the well-known vertical slit typical of vipers when threatened. *Scales* Dorsals smooth and in 17 or 19 rows at midbody; scales without any pits; ventrals 143–176; subcaudals 44–72 and paired; anal divided. *Colouration* Dorsum colour varies largely from yellowish-grey to reddish-brown, to dark brown or black; irregular, faint marks, usually in form of dark-edged, light brown ovals or bars on body; some with longitudinal stripes in mid-dorsal and sides of body; head with symmetrical, longitudinal markings on top and sides, including streak through eye; venter speckled with brown dots. **HABITAT AND HABITS** Occurs in leaf litter-rich montane forests (mainly in hills), but also encountered closer to human habitats. Prefers locations close to water. Displays both nocturnal and diurnal behaviours. Mostly terrestrial, but may climb bushes. Generally feeds on hard-bodied reptiles such as skinks and snakes, but soft-bodied animals such as geckos and frogs are also taken. Prey subdued by venom and constriction. Ovoviviparous, giving birth to up to 10 young at a time. Can reproduce a few times in a year. **VENOM** Rear-fanged (ophisthoglyphous) snake. Symptoms of bite include localized mild pain lasting for short periods, and sleepiness.

Adult colourations.

Indo-Chinese Sand Snake ■ *Psammophis indochinensis* TL 90cm
(*Bahasa Indonesia* Ular Pasir Asia)

DESCRIPTION *Physique* Body slender and long; head oval with distinct neck; eyes large with round pupils. *Scales* Dorsals smooth and in 17 or 19 rows at midbody; ventrals 150–173; subcaudals 75–85 and paired; anal divided. *Colouration* Dorsum light brown or olive-brown with 4 dark brown longitudinal stripes from head along body; these lines are edged with thin black line; some with very faint markings; venter yellowish with darker line at outer margins. **HABITAT AND HABITS** Mostly inhabits open habitats, including agricultural systems and home gardens in middle elevations. Displays both terrestrial and arboreal behaviours. Diurnal. Feeds on amphibians, reptiles (including other snakes) and mammals (mostly rodents). Oviparous. **VENOM** Non-venomous snake. However, swelling may occur after bites.

Adult colouration.

ELAPIDAE
TERRESTRIAL ELAPID SNAKES

The Elapidae family currently contains nearly 375 members, including the true sea snakes and sea kraits, which are discussed in a separate section (see p. 142). The terrestrial members are widespread in the world, and occupy all continents except Europe and Antarctica.

Elapids vary greatly in body size and form, and range from minute burrowing forms to very large-bodied terrestrial ones. In morphology they are similar to colubrids, and both groups have large, symmetrically arranged head shields, broad ventral scales and fewer than 30 midbody scale rows. However, unlike colubrids, elapids lack a loreal scale between the nostril and the eye.

Elapids occupy all habitats (fossorial, aquatic, arboreal and terrestrial), and mostly feed on vertebrate prey (with a few species specializing in reptile eggs). Ophiophagy, or feeding on other snakes, is common among these members. The group includes both oviparous and ovoviviparous species.

Many elapids are potentially deadly venomous snakes. All have a pair of 'proteroglyphous fangs', where the enlarged and hollow fangs are located at the front of the mouth. When the mouth is closed, the fangs fit into grooved slots in the mouth floor. Due to forwards-facing holes at the tips of their fangs, a few species are capable of spraying their venom using pressure (the group known as the spitting cobras).

Five species of terrestrial elapid are known from Bali.

Occasionally growing to more than 5m in length, the King Cobra is the largest venomous snake species in the word.

Malayan Krait ■ *Bungarus candidus* TL 150cm
(*Bahasa Indonesia* Ular Weling; *Balinese* Lélipi Poleng, Lélipi Sélém Bali, Lélipi Ngis)

DESCRIPTION *Physique* Body triangular in cross-section; head small and with slightly distinct neck; eyes small and completely black; tail short and blunt. *Scales* Dorsals shiny, smooth and in 15 or 17 rows at midbody; vertebral scales significantly larger than adjoining body scales, and hexagonal in shape; ventrals 194–237; subcaudals 37–56 and single; anal undivided. *Colouration* Dorsum black, dark brown or copper-brown, with some individuals having 20–35 broad white cross-bands along body. These bands usually widen up at the flanks, and are more prominent on juveniles than adults. Faint light chevron on neck of some individuals; lips white.

HABITAT AND HABITS Mostly occurs in lowlands, and found in forested habitats, plantations and semiurban areas, mainly close to water. Nocturnal and terrestrial. During day, hides and rests in termite mounds and rodent holes, underneath and among stones, or beneath heaps of coconut husks and leaves. Feeds on other snakes, amphibians, lizards and small mammals. Oviparous, with clutch size of up to 10 eggs. **VENOM** Front fixed-fanged (proteroglyphous), highly venomous snake. Symptoms of snake bite may include headache, nausea, vomiting, abdominal pain, diarrhoea, dizziness, collapse or convulsions, and flaccid paralysis is a major clinical effect. Bites potentially fatal if untreated.

Adult colourations.

Adult colourations, Malayan Krait.

Banded Krait ■ *Bungarus fasciatus* TL 230cm
(*Bahasa Indonesia* Ular Welang; *Balinese* Lélipi Poleng Kuning)

DESCRIPTION *Physique* Body triangular in cross-section, with raised vertebral region; head small with slightly distinct neck; eyes small and completely black; tail short and blunt. *Scales* Dorsals shiny, smooth and in 15 rows at midbody; vertebral scales significantly larger than adjoining body scales and hexagonal in shape; ventrals 200–236; subcaudals 23–39 and single; anal undivided. *Colouration* Dorsum bright or pale yellow, with 14–32 wide black cross-bands (more or less same width as yellow interspaces) on body and 2–5 on tail; head black with yellow 'V'-shaped mark; lips yellow. **HABITAT AND HABITS** Mostly occurs in lowlands, but may occasionally be encountered in mountains. Found in forested habitats, plantations and semiurban areas, mainly close to water. Nocturnal and terrestrial. During day, hides and rests in termite mounds and rodent holes, underneath and among stones, or beneath heaps of coconut husks and leaves. Commonly balls or forms body into coils when disturbed. Caudal luring shown by juveniles. Feeds on other snakes, frogs, lizards, fish, reptile eggs and even small mammals. Oviparous, with clutch size of up to 14 eggs. **VENOM** Front fixed-fanged (proteroglyphous), highly venomous snake. Symptoms of snake bite may include headache, nausea, vomiting, abdominal pain, diarrhoea, dizziness, collapse or convulsions, and flaccid paralysis is a major clinical effect. Bites potentially fatal if untreated.

Adult colourations.

Malayan Striped Coral Snake ■ *Calliophis intestinalis* TL 70cm
(*Bahasa Indonesia* Ular Cabe)

DESCRIPTION *Physique* Body slender and long; head small with indistinct neck; eyes small and completely black; tail short and blunt. *Scales* Dorsals smooth and in 13 or 15 rows at midbody; ventrals 197–273; subcaudals 15–33 and divided; anal undivided. *Colouration* Subspecies *C. i. intestinalis*, occurring in Bali, has blackish dorsum, narrow white or yellowish longitudinal lines along vertebral and lower flank regions, and brown stripe in middle (on mid-flanks); venter reddish with broad white-and-black transverse bands; black forehead with 'Y'-shaped light mark; other subspecies have distinct colourations. **HABITAT AND HABITS** Mostly occurs in lowlands and midlands, and found in well-forested as well as more open habitats. Terrestrial and semifossorial. Nocturnal, spending day under logs, rocks and leaf litter. When disturbed, coils and flattens body, and often hides head underneath body. Also curls tail over back and exposes banded ventral aspect of body. Feeds on other, smaller snakes. Oviparous, with clutch size of up to 3 eggs. **VENOM** Front fixed-fanged (proteroglyphous), venomous snake. Mild local reactions recorded after bites, but there are insufficient clinical reports. Generally considered as not dangerous due to small size.

Adult colouration.

Southern Indonesian Spitting Cobra ■ *Naja sputatrix* TL 110cm
(*Bahasa Indonesia* Ular Sendok, Ular Kobra Penyembur; *Balinese* Lélipi Sendok, Lélipi Who; *English* Equatorial Spitting Cobra)

DESCRIPTION *Physique* Body robust and cylindrical; head large, depressed and broad behind, with slightly distinct neck; eyes moderate with round pupils; tail round and tapering. *Scales* Dorsals smooth and in 17, 19, 21, 23 or 25 rows at midbody (some island populations have more than others); ventrals 160–188; subcaudals 40–56 and divided; anal undivided. *Colouration* Highly variable, and also shows geographic variations. Dorsum blackish, greyish, brownish, yellowish or orangey; throat patterning faint and not well defined; hood may be unpatterned or contain a spectacle-shaped or horseshoe mark; venter whitish, patterned with darker spots in some individuals. **HABITAT AND HABITS** Mostly occurs in lowlands, and frequently encountered near rock outcrops and wetland habitats, including paddy fields and plantations. Displays both nocturnal and diurnal, and terrestrial and arboreal behaviours. When disturbed and/or threatened, may rear, spread hood and spit venom or bite. Feeds mostly on small mammals (mainly rodents), but also takes other snakes, amphibians and lizards. Oviparous, with clutch size of up to 26 eggs. **VENOM** Front fixed-fanged (proteroglyphous), highly venomous snake. Marked local effects of snake bite include pain, severe swelling, bruising, blistering and necrosis. Systemic effects may include nausea, abdominal pain, diarrhoea, collapse and flaccid paralysis. Bites potentially fatal if untreated. Venom sprayed on to eyes may result in pain and temporary blindness.

Adult colouration.

Adult colourations, Southern Indonesian Spitting Cobra.

King Cobra ■ *Ophiophagus hannah* TL 300cm
(*Bahasa Indonesia* King Kobra; *Balinese* Lelipi Selang Bebek; *English* Hamadryad)

DESCRIPTION *Physique* Body robust and very long (slender in juveniles); head large, depressed and broad behind, with slightly distinct neck; eyes large with round pupils; tail short. *Scales* Dorsals smooth and in 15 rows at midbody; vertebral row and outer 2 lateral rows larger than others; ventrals 215–270; subcaudals 74–125, front ones undivided; anal undivided. *Colouration* Dorsum brownish, blackish or yellowish, and ranging from distinctly patterned with lighter or darker narrow bands, to completely unpatterned; hood has chevron marks in some individuals; juveniles usually have white or yellow cross-bars, which are chevron shaped on front of body but straighten out on latter part; throat yellowish, with some individuals having darker cross-bands towards bottom. **HABITAT AND HABITS** Mostly occupies forested areas and thick plantations with vegetation cover. Diurnal and occasionally nocturnal. Terrestrial as adult, but juveniles are more arboreal. When disturbed and/or threatened, may rear, spread hood and bite. Feeds on other snakes and large lizards such as monitor lizards. Builds nest using fallen leaves (the only snake species known to build a nest). Oviparous, with clutch size of 20–51 eggs. Females guard eggs. **VENOM** Front fixed-fanged (proteroglyphous), highly venomous snake. Marked local effects of snake bite include pain, severe swelling, bruising, blistering and necrosis. Systemic effects may include nausea, abdominal pain, diarrhoea, collapse and flaccid paralysis. Bites potentially fatal if untreated.

Adult (left) *and hood* (right).

■ TRUE SEA SNAKES & SEA KRAITS ■

ELAPIDAE
TRUE SEA SNAKES AND SEA KRAITS

About 70 elapid species have adapted to life in salt water. Most require warm waters, so are found in tropical and subtropical waters of the Indian and Pacific Oceans, with only a handful of records from the Atlantic Ocean. Two major radiations have taken place in the Asian-Australian region, the true viviparous 'sea snakes' and the oviparous 'sea kraits'. True sea snakes have flattened bodies and extremely reduced ventral scales, which makes it impossible for them to move on land, while sea kraits are truly amphibious and move easily on land.

Both groups show several unique adaptations to life in the sea. They have vertically flattened, paddle-like tails for propulsion (not found in any other snake group, including other freshwater and brackish-water species), dorsally positioned nostrils each with a muscular valve that closes when they dive, salt regulating glands (including specialized excretory glands on the base of the tongue), and a single lung that extends nearly the full length of the body, which enables them to do long dives (sometimes lasting 2 hours or even longer). Although sea snakes are able to absorb a large portion of their oxygen requirements from sea water through the skin, all are air breathers and therefore need to rise to the surface to breathe. The Olive Sea Snake *Aipysurus laevis* is reported to have a light receptor on the tip of its tail. This may allow sheltering snakes to keep their tail paddles retracted and out of reach of predators.

Some members of this group have generalist diets, but most have highly specialized preferences, feeding almost entirely on eels, gobies and catfish-like prey, or fish eggs. There are also records of crustaceans and molluscs being part of sea-snake diets. A few species have extremely small heads and very slender forebodies that enable them to probe crevices in search of food. All true sea snakes are ovoviviparous, and give birth to live young in the ocean. Sea kraits lay eggs on land.

Most species inhabit shallow waters along coasts, around islands and coral reefs, and in river mouths. An exception is the Yellow-bellied Sea Snake *Hydrophis platurus*, which lives in open oceans, so is one of the most widely distributed snakes in the world.

As a general rule all sea snakes must be regarded as dangerously venomous and should be handled with great caution. Some species are inoffensive and would only bite under provocation, but others are much quicker to defend themselves when threatened. Sea snakes are quite curious creatures and sometimes approach people in the water. Avoid interacting with one if it approaches you, and wait patiently until it moves off.

Two species of sea krait (*Laticauda colubrina* and *L. laticaudata*) are seen on southern coasts of Bali, where they come ashore in some places. Another species (*L. semifasciata*) has been recently seen in offshore waters of Bali, and also known from nearby islands. However, no proper assessment of the true viviparous sea snakes of Bali has been conducted. The following key is for species probably occurring in the area based on records from other nearby islands, personal observations and expert opinion.

KEY TO IDENTIFYING MARINE ELAPIDS OF BALI

For confirmation of species identity, more technical keys that incorporate detailed scale counts, tooth counts and internal characters should be consulted.

1.	Nasals separated by internasals (a.), so nostrils placed laterally; body mostly cylindrical; can actively move on land.	Sea kraits (go to 2.)
	Nasals not separated by internasals (b.), so nostrils placed dorsally; body mostly laterally flattened; does not actively move on land.	Viviparous sea snakes (go to 3.)

a.

b.

2.	Upper lip yellowish; rostral scale not divided; 21–25 scale rows at midbody; body bands usually narrower or equal in width to the gaps between the bands	Yellow-lipped Sea Krait *Laticauda colubrina*
	Upper lip brownish or blackish; rostral scale not divided; 19 scale rows at midbody; body bands usually narrower or equal in width to the gaps between the bands	Brown-lipped Sea Krait *Laticauda laticaudata*
	Upper lip brownish; rostral scale divided horizontally (a smaller upper part between nasals and a wider bottom part); 21 or 23 scale rows at midbody; body bands usually wider than the gaps between the bands	Chinese Sea Krait *Laticauda semifasciata*

Yellow-lipped Sea Krait   *Brown-lipped Sea Krait*   *Chinese Sea Krait*

3.	Dorsal side of body black and bottom half yellowish, with colours sharply delineated (rarely all-yellow or all-black individuals occur); head elongated, with flattened, bill-like snout; very wide gape (c.); SVL ♀66cm, ♂56cm.	Yellow-bellied Sea Snake *Hydrophis platurus*
	Not as above.	Go to 4.

c.   Yellow-bellied Sea Snake

4.	Some scales on head raised.	Go to 5.
	No scales on head raised.	Go to 6.

5.	Free edges of supraoculars raised into blunt, spinous horn-like structures (d.); head small relative to body; SVL about 90cm.	Horned Sea Snake *Hydrophis peronii*
	Free edges of head shields thickened and may appear raised, especially around snout; rostral divided into 4 or 5 scales; pair of elongated shields separate nasals (e.); short and stout body; SVL ♀72cm, ♂67cm.	Anomalous Sea Snake *Hydrophis anomalus*

d.   Horned Sea Snake

e.   Anomalous Sea Snake

6.	Ventrals large, each at least 3 times as broad as adjacent body scales (f.), at least on forebody.	Go to 7.
	Belly scales (ventrals) small, either indistinguishable (g.), or each not more than 2 times as broad as adjacent body scales (h.).	Go to 9.

f.

g.

h.

7.	Ventrals half as broad as body anteriorly (i.), narrowing gradually until they are not twice as broad as adjacent scales posteriorly (j.); usually distinctly counter shaded with grey or bluish-grey dorsally and whitish ventrally; distinct plate-like chin shields, with anterior pair smaller than posterior pair; SVL ♀83cm, ♂74cm.	Viperine Sea Snake *Hydrophis viperinus*
	Ventrals more or less equal in width and more than twice as broad as adjacent body scales.	Go to 8.

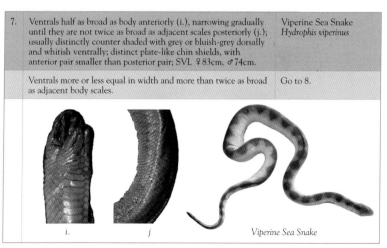

i.

j

Viperine Sea Snake

| 8. | Head scales large and symmetrically arranged (k.); lip scales not divided horizontally; dorsal cream to salmon coloured, with series of broad, often irregular dark cross-bands; SVL about 85cm. | Spine-tailed Sea Snake *Aipysurus eydouxii* |
| | Head scales more or less broken up into smaller scales; some lip scales divided horizontally (l.); colouration extremely variable; SVL ♀86cm, ♂79cm (max. recorded 200cm). | Olive Sea Snake *Aipysurus laevis* |

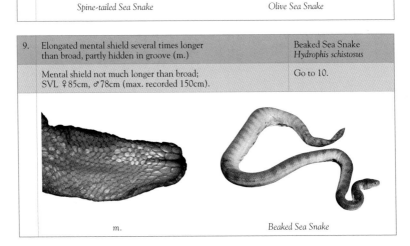

k.

l.

Spine-tailed Sea Snake

Olive Sea Snake

| 9. | Elongated mental shield several times longer than broad, partly hidden in groove (m.) | Beaked Sea Snake *Hydrophis schistosus* |
| | Mental shield not much longer than broad; SVL ♀85cm, ♂78cm (max. recorded 150cm). | Go to 10. |

m.

Beaked Sea Snake

10.	Very thin forebody with very small head, and very robust and laterally compressed hindbody.	Go to 11.
	Entire body of fairly uniform thickness, or head and neck slightly slender compared to rest of body.	Go to 12.

11.	Ventrals divided by longitudinal groove towards back of body; SVL ♀93cm, ♂87cm.	Slender Sea Snake *Microcephalophis gracilis*
	Ventrals not divided by longitudinal groove towards back of body; SVL about 90cm.	Black-headed Sea Snake *Hydrophis atriceps*

Slender Sea Snake   Black-headed Sea Snake

12.	Robust body with large head and thick neck; belly scales (ventrals) very small and divided into 2 strongly overlapping rows forming distinct ventral keel (except on throat) (n.); with or without large dark dorsal blotches sometimes alternating with narrow bands; SVL ♀141cm, ♂10 cm (max. recorded 200cm).	Stoke's Sea Snake *Hydrophis stokesii*
	Not as above.	Go to 13.

n.   Stoke's Sea Snake adult and juvenile.

13.	Ventrals very small and difficult to distinguish (g.); dorsal scales non-overlapping (juxtaposed) and small, while lowermost 3 or 4 rows larger than others and with round central tubercles (o.); stout body; SVL ♀83cm, ♂72cm. **Note** Males of this species as well as several others may develop spinous tubercles on scales during breeding season, as shown in image.	Spine-bellied Sea Snake *Hydrophis curtus*
	Not as above.	Go to 14.

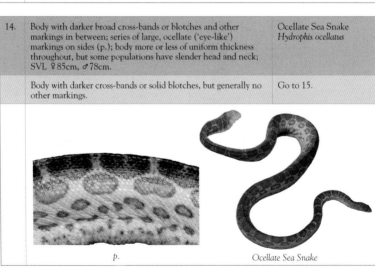

o.   Spine-bellied Sea Snake colour variations

14.	Body with darker broad cross-bands or blotches and other markings in between; series of large, ocellate ('eye-like') markings on sides (p.); body more or less of uniform thickness throughout, but some populations have slender head and neck; SVL ♀85cm, ♂78cm.	Ocellate Sea Snake *Hydrophis ocellatus*
	Body with darker cross-bands or solid blotches, but generally no other markings.	Go to 15.

p.   Ocellate Sea Snake

15.	Interspaces between bands 2–4 times broader than bands posteriorly; primary body colour yellowish or yellowish-green; head of adults almost entirely yellowish, but blackish in juveniles with yellow horseshoe mark above; SVL ♀171cm, ♂148cm (max. recorded 275cm).	Yellow Sea Snake *Hydrophis spiralis*
	Interspaces between bands narrower than bands, at least on dorsal body.	Go to 16.

Yellow Sea Snake colour variations

16.	Parietals not touching postoculars; body and tail with 40–60 blackish blotches or transverse bands; head black or dark grey with or without light streak behind eye.	Dwarf Sea Snake *Hydrophis caerulescens*
	Parietals touching postoculars.	Go to 17.

Dwarf Sea Snake

17.	Costals with short bi- or tri-dentate keel (or rarely smooth); small, wedge-shaped scales (cuneates) along border of mouth between infralabials; interspaces between bands narrower than bands posteriorly; primary colour cream/yellow/white, bands black-bluish; head of adult olive or yellowish with some individuals with horseshoe marking on head, especially in juveniles; SVL ♀175cm, ♂137cm.	Annulated Sea Snake *Hydrophis cyanocinctus*
	Costals with round central tubercle or short median keel, or smooth; colouration not as above.	Go to 18.

Annulated Sea Snake

18.	Body and tail with less than 45 blackish cross-bands that are usually widest on dorsal and ventral surfaces; head black with conspicuous yellow marks on snout and behind eyes in most individuals; primary body colour drab olive-grey; SVL about 90cm.	Cogger's Sea Snake *Hydrophis coggeri*
	Body and tail has more than 45 blackish cross-bands that are widest on dorsal and ventral surfaces.	Go to 19.

Cogger's Sea Snake

19.	Primary body colour cream-yellow throughout; body and tail has 50–70 black cross-bands that are twice as broad as interspaces between; bands rarely taper towards venter; head black with conspicuous yellow marks; SVL ♀ 111cm, ♂ 102cm.	Black-banded Sea Snake *Hydrophis melanosoma*
	Primary body colour cream-yellow above, much paler below; body and tail with at least 52 darker cross-bands that are twice as broad as interspaces between; bands taper towards venter; head black flecked with olive, with traces of horseshoe-shaped yellow mark on prefrontals; SVL about 90cm.	Belcher's Sea Snake *Hydrophis belcheri*

Black-banded Sea Snake

Belcher's Sea Snake

HOMALOPSIDAE
MANGROVE SNAKES

Members of this family were previously placed under the loosely bound Colubridae. They are now in their own family of nearly 55 members, distributed in South and Southeast Asia, New Guinea and Australia.

Homalopsids generally have robust bodies and round, tapering tails (unlike the flattened, paddle-like tails of true sea snakes and sea kraits). Their valvular nostrils and eyes are placed more dorsally on the head, enabling them to breathe and see at the surface of the water without exposing much of the head or body. Their ventral scales are broad, enabling them to move on land.

These snakes are aquatic or semiaquatic and largely nocturnal. They inhabit a wide range of freshwater habitats, as well as coastal ones including mangroves, mudflats and estuaries. While most species feed on goby-like fish, a few are specialist feeders on crustaceans. This group also includes two species of snake (Gerard's Water Snake *Gerarda prevostiana* and White-bellied Mangrove Snake *Fordonia leucobalia*) that are known to dismantle their prey rather than swallowing it whole. All members are ovoviviparous.

Homalapsids are rear-fanged venomous snakes with an ophisthoglyphous fang arrangement, but no species is considered dangerous to humans, and there are no fatal bite records from any homolapsid.

Two homolapsid species are known from Bali.

Dorsally located eyes are characteristics of homalopsids.

Schneider's Bockadam ■ *Cerberus schneiderii* TL 70cm
(*Bahasa Indonesia* Ular Tambak; *English* Schneider's Dog-faced Water Snake)

DESCRIPTION *Physique* Body stout; snout broadly rounded and with prominent lower jaw; eyes small and bulbous; eyes and valvular nostrils located towards top of head; tail short and slightly compressed. *Scales* Imbricate plate-like scales on forehead; first supralabial does not contact loreal; last supralabial horizontally divided; dorsals strongly keeled and in 23 or (rarely) 25 rows around midbody; ventrals 122–160; subcaudals 49–72 and paired; anal divided. *Colouration* Dorsum greyish-green, dark grey or reddish, with series of inconspicuous dark, irregular bars or blotches; conspicuous dark line along sides of head, across eyes; lowermost costals have black-and-cream markings that alternate; venter cream with diffused black markings. **HABITAT AND HABITS** Occurs in mangroves, estuaries and mudflats, and occasionally in fresh water, paddy fields and slow-moving, shallow streams, and on sheltered sandy coasts. Also found in ditches and small, man-made canals in coastal towns. Moves well on land and frequently sidewinds on muddy substrates when approached. Largely nocturnal, but occasionally seen during day and at dusk. Spends the day down crab burrows or among mangrove roots. Both a sit-and-wait species and an active forager, feeding on fish, crustaceans and frogs. Feeding aggregations of many closely spaced individuals can occur. Ovoviviparous, producing up to 30 young, sometimes multiple times in a year. When provoked strikes repeatedly, and emits foul odour from anal glands. **VENOM** Rear-fanged (ophisthoglyphous) snake. No systemic effects from bites known.

Adult colouration and ventral aspects (inset).

Olive Water Snake ■ *Hypsiscopus plumbea* TL 40cm
(*Bahasa Indonesia* Ular Lumpur; *English* Boie's Mud Snake, Plumbeous Water Snake)

DESCRIPTION *Physique* Body short, robust and cylindrical; head short with round snout and indistinct neck; tail short; eyes small and beady with round pupils, and located dorsally. *Scales* Dorsals smooth and in 19 rows around midbody; ventrals 112–139; subcaudals 22–46 and paired; anal divided. *Colouration* Dorsum olive-green, greyish-green or brownish-green, with most body scales having darker edges (the species name '*plumbea*' refers to its lead-like

colouration); lips and venter cream or yellowish; some individuals mottled with black markings, or with dark line along dorsum; anal and subcaudal scales have dark median line. **HABITAT AND HABITS** Occurs in freshwater bodies, including ditches and paddy fields; also ventures into brackish water habitats. Largely nocturnal, but occasionally seen during day and at dusk. Feeds on crustaceans, amphibians and their eggs, and fish. Ovoviviparous, producing up to 30 young. **VENOM** Rear-fanged (ophisthoglyphous) snake. Minor local swelling and pain known after a bite.

Adult colourations.

PAREIDAE
ORIENTAL SLUG-EATERS

Members of this family were once placed under the loosely bound family Colubridae. They are now in their own family of close to 25 members, widely distributed in tropical and subtropical regions of Southeast Asia.

Pareatids have smallish, thinly built, slightly vertically compressed bodies. The head is large with a distinct neck, large eyes and a blunt snout. The scales on the chin are large and overlapping. Most species have brownish cryptic colouration, which provides camouflage among vegetation.

These snakes are arboreal and largely nocturnal. They inhabit a wide range of vegetated habitats, including home gardens. They are specialized mollusc eaters and prey on a wide variety of snails and slugs. However, some species occasionally also feed on invertebrates and small vertebrates. A few family members, such as the Blunt-headed Slug Snake *Aplopeltura boa*, are known to feign death, employing strategies such as immobility and rolled body when disturbed.

Pareatids are non-venomous snakes and are not considered dangerous to humans.

A single species of pareatid snake is known from Bali.

Bali's only pareatid species, the Keeled Slug-eater.

Keeled Slug-eater ■ *Pareas carinatus* TL 50cm
(*Bahasa Indonesia* Ular Siput)

DESCRIPTION *Physique* Body slender and laterally compressed; head broad and round, with very short and blunt snout and distinct neck; eyes very large with vertically elliptical pupils. *Scales* Chin shields asymmetrical and without mental groove; dorsals enlarged in 2–3 median rows (and weekly keeled), and in 15 or 17 rows at midbody; ventrals 158–207; subcaudals 53–111 and paired; anal undivided. *Colouration* Dorsum reddish-brown,

yellowish or olive-brown, with weak darker cross-bands or transverse spots on body; these marks are more prominent on front part of body; two dark streaks behind each eye, the top one more prominent and making prominent 'X'-shaped mark on nape; darker spot on each end of ventrals forms 2 longitudinal lines along venter; iris orange-red. **HABITAT AND HABITS** Nocturnal, sluggish and arboreal, and often encountered on low vegetation. Occurs in wide variety of habitats from lowlands to submontane areas, often close to water. Feeds on snails and slugs. When feeding on snails, bites into a snail's body and pulls it out with alternate retraction movements of mandibles while upper jaw rests on outside of shell. Oviparous, with clutch size of up to 8 eggs. **VENOM** Non-venomous snake.

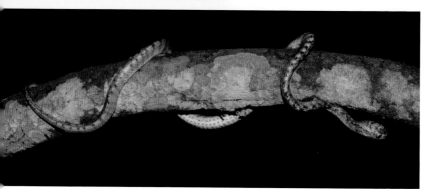

Close-up of head (top) *and adult colouration.*

VIPERIDAE
VIPERS AND PIT VIPERS

The Viperidae family currently comprises nearly 355 members, which are widespread in the world and occupy all continents except Antarctica and Australia.

Viperids vary largely in body size and form, ranging from minute arboreal species to very large-bodied terrestrial ones. Morphologically they are distinct, with large, triangular heads covered with small scales (rather than large and symmetrically arranged head shields), and relatively stout bodies covered with mostly keeled scales. Pit vipers have a specialized organ named the 'loreal pit' between the eye and the nostril on each side of the face, which contains infrared receptors that allow the snake to detect warm-blooded prey through thermal signals.

Viperids occupy all habitats (semifossorial, semiaquatic, arboreal and terrestrial) and mostly feed on warm-blooded prey such as birds and mammals. Most members are ovoviviparous, but a few lay eggs.

Many viperids are potentially deadly venomous snakes and have a pair of 'solenoglyphous fangs', which are hinged at the front of the mouth and fold back against the palate when not in use. An elongated bone functions as a lever to move the fang when the mouth is opened. Most species have haemotoxic venom that affects the blood.

A single species of pit viper is known from Bali, and the highly venomous Eastern Russell's Viper *Daboia siamensis* may possibly occur in the island. The taxonomy of the pit vipers is quite unstable and their scientific names are subjected to regular changes.

Eastern Russell's Viper may be identified by the triangular head with a 'V'-shaped mark and the three series of chain-like dark brown spots that run the length of the body. There are no confirmed records of this species from Bali.

Lesser Sundas White-lipped Pit Viper

▪ *Trimeresurus albolabris* TL ♀ 85m, ♂ 60m
(*Bahasa Indonesia* Ular Mati Ekor, Ular Majapahit, Ular Bungka Laut, Ular Hijau; *Balinese* Lélipi Gadang Sugém, Lélipi Gadang Ikut Barak, Lélipi Tabié)

DESCRIPTION *Physique* Body stout and medium in size; head large, triangular, covered with small and irregular scales, and with distinct neck; eyes large with vertical pupils; heat-sensitive loreal pit between nostril and eye on either side; tail short and prehensile. *Scales* Dorsals keeled (lowermost laterals unkeeled), and in 19, 21 or 23 rows at midbody; ventrals 153–176; subcaudals 48–81 and paired (more in males than females); anal undivided. *Colouration* Dorsum green (various shades), but individuals may be yellowish or bluish in other parts of the range; lips and throat light green or yellowish, with occasional white stripe at margin of two colour zones; some individuals have transverse dark bands on body; tail with rust or reddish colour streak. **HABITAT AND HABITS** Mostly occurs in lowlands, and commonly found in forested areas, bamboo thickets, plantations (including around paddy fields) and semiurban areas. Arboreal, and prehensile tail is used to cling to branches while moving on vegetation. However, also displays terrestrial behaviours. Slow moving and nocturnal. Feeds on frogs, lizards, birds and mammals (mostly rodents). Also cannibalistic. Some individuals vibrate their tails when irritated. Ovoviviparous, giving birth to up to 17 young. **VENOM** Front-articulated fanged (solenoglyphous), venomous snake. Bites known to cause severe pain, bleeding and swelling at bite site, and rarely death. Should be considered a dangerous species.

Loreal pit between eye and nostril.

Adult colourations.

GERRHOPILIDAE AND TYPHLOPIDAE
BLIND SNAKES

These two families include nearly 300 primitive fossorial snakes. The two groups are found in parts of Asia, Africa, America, south-east Europe and Australia, as well as on some Pacific islands.

All blind snakes are fossorial, and their earthworm-like bodies are well adapted to a burrowing lifestyle. However, occasionally individuals are found climbing trees and even within epiphytes. Their bodies are small (minute in some) and of uniform thickness, and covered with non-overlapping, shiny scales. The head is not well defined, blunt and with a large rostral in the form of a shield. The eye is covered with a transparent scale to prevent damage while burrowing. The tail is short and ends in a spine in some species. Females are significantly larger than males in most species.

Due to the fossorial and cryptic lifestyle of these snakes, very little is known about their biology. They are insectivorous, feeding on eggs, pupae and adults of fossorial insects, mostly termites and ants. Some live inside or very close to the nests of these social insects. They are non-venomous, but may release foul-smelling excretions from the cloacal glands if handled. At least one species occurs in all-female parthenogenetic populations.

Published literature and online databases list two species of typhlopid and one gerrhopilid from Bali and Nusa Penida. Several other species are likely to occur in the islands.

Note Identification of blind snakes to species level is very difficult and requires careful observation of the head scales and other body characteristics, so it can be challenging for a non-specialist. For confirmation of species identity, refer to more technical sources that incorporate scale counts and internal characteristics.

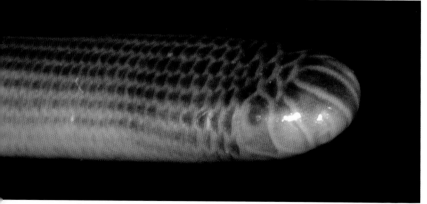

A not well-defined head with a blunt and large rostral, and eyes each covered with a transparent scale, are characteristics of blind snakes. This is a close-up of the head of a Lesser Sundas Blind Snake.

Lesser Sundas Blind Snake ■ *Sundatyphlops polygrammicus* TL 25cm
(*Bahasa Indonesia* Ular Buta Sunda Kecil)

DESCRIPTION *Physique* Body slender, cylindrical; head indistinct from neck; snout rounded; nostrils placed laterally; eye distinct; caudal spine present. *Scales* Rostral almost extends to eye; nasal mostly incompletely divided; supralabials 4 with 3rd beneath eye; postoculars 2–3; no suboculars; inferior nasal cleft starts from either junction of 1st and 2nd supralabial, or from 2nd supralabial; head scales larger than dorsals; dorsals smooth and highly polished; scale rows 22; ventrals very small; subcaudals 10–19. *Colouration* Dorsum dark or olive with alternating dark and light longitudinal lines along body; venter lighter, with some scales with darker spots; lips whitish; underside of tail has darker spots. **HABITAT AND HABITS** Mostly found in mountainous regions. Nocturnal and fossorial, or active on the surface, especially after rains. Mainly feeds on pupae and larvae of ants. Oviparous, with clutch size of up to 7 eggs. **VENOM** Non-venomous snake.

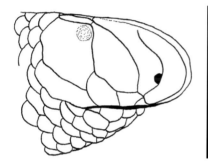 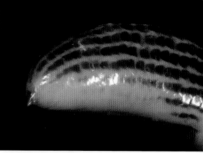

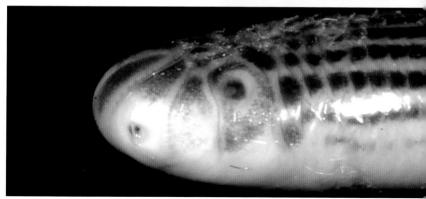

Head scalation (bottom) and tail-tip (top right).

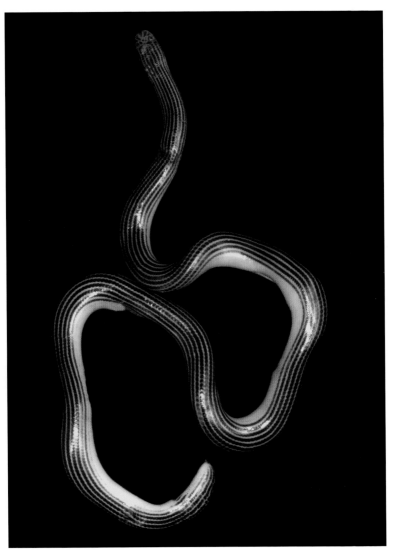

Adult, Lesser Sundas Blind Snake (preserved specimen WAM.R98715).

Black Blind Snake ■ *Gerrhopilus ater* TL 15cm
(*Bahasa Indonesia* Ular Buta Hitam)

DESCRIPTION *Physique* Body slender and long; total length about 68 times diameter of body; head indistinct from neck; snout rounded; nostrils placed laterally; eye distinct; tail twice as long as wide; caudal spine present. *Scales* Rostral elongated, oval and extends to level of eyes; numerous distinct papillae-like structures (sebaceous glands) cover head shields; supralabials 4; nasals usually in contact behind rostral or overlapping one another; preocular in contact with subocular or 2nd and 3rd supralabials; subocular often present; scale rows 18. *Colouration* Dorsum dark brown or black; venter reddish-brown; base of tail and chin cream.
HABITAT AND HABITS Probably a fossorial species, living under leaf litter and loose soil. Possibly feeds on arthropods and earthworms. Reproductive habits unstudied. **VENOM** Non-venomous snake.

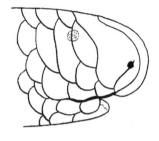

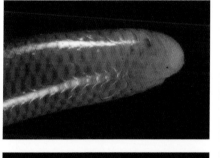

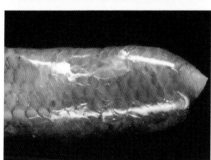

Head scalation (top and middle), *tail-tip* (bottom left), *and adult* (*preserved specimen RMNH.RENA 6297*) (bottom right).

Flowerpot Snake ■ *Indotyphlops braminus* TL 12cm
(*Bahasa Indonesia* Ular Kawat, Ular Buta Brahminy)

DESCRIPTION *Physique* Body slender, cylindrical; total length about 30–45 times diameter of body; head indistinct from neck; snout evenly rounded; nostrils placed laterally; eye distinct; caudal spine present. *Scales* Supralabials 4 with 3rd beneath eye; preocular 1 and in touch with supralabials; postocular 1; no suboculars; inferior nasal cleft starts from preocular; nasal completely divided; head scales larger than dorsals, and with noticeable but minute tubercules; dorsals smooth and highly polished, in 20 rows at midbody; ventrals 280–320, very small; subcaudals 8–15. *Colouration* Dorsum uniform black, dark grey, dark brown or purplish-brown; venter lighter; tail-tip and snout paler or cream coloured. **HABITAT AND HABITS** The most widely distributed snake species in the world. Encountered in human habitations as well as in lightly forested areas, from sea level to at least 2,000m asl. Nocturnal and fossorial, or active on the surface, especially after rains. Can sometimes be found beneath bark, in rotten wood or in termite runs in trees some distance above the ground. Feeds on termites, ants and their larvae, and earthworms. Known to decapitate termite prey and consume only thorax and abdomen. Parthenogenetic, and only females are known. Lays 1–8 eggs. **VENOM** Non-venomous snake.

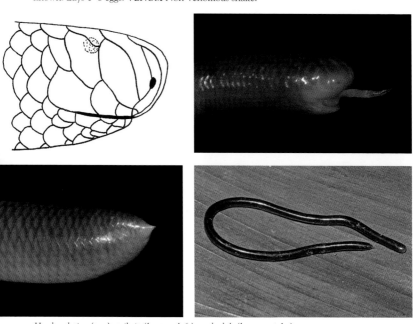

Head scalation (top), *tail-tip* (bottom left), *and adult* (bottom right).

CROCODYLIDAE
TRUE CROCODILES

This is the largest family of the order Crocodilia (the others being the Aligatoridae, which includes alligators and caimans, and the Gavialidae, which includes a sole member, the Gharial). The family comprises 12 species of true crocodiles (*Crocodylus* spp.), the False Gharial *Tomistoma schlegelii*, two species of African Slender-snouted Crocodiles *Mecistops* spp., and the African Dwarf Crocodile *Osteolaemus tetraspis*. They are naturally distributed in tropical parts of Asia, Africa and the Americas.

True crocodiles have narrow, long heads with both jaws being of the same width (alligatorids have a wider upper jaw). The large 4th tooth in the lower jaw thus fits into a constriction in the upper jaw when crocodiles close their jaws (in alligatorids, all teeth in the lower jaw fit into small depressions in the upper jaw). The body is heavily armoured with bony plates. While most species are found in fresh water, some inhabit brackish water and even occasionally traverse open seas.

Crocodilians are opportunistic and generalist feeders, with their prey ranging from molluscs and crustaceans, to large mammals. All species are oviparous and show temperature dependent sex determination (TSD), where the sex of the offspring is determined by the incubation temperature rather than by genetics. In most species the mother carries the newly hatched babies to water and guards them, sometimes for up to several months.

One species occurs in Bali and around Nusa islands.

Reaching more than 5m in length, the male Saltwater Crocodile is the largest living reptile species.

Saltwater Crocodile ▪ *Crocodylus porosus* TL ♀ 5m, ♂ 3m
(*Bahasa Indonesia* Buaya Muara)

DESCRIPTION *Physique* The largest living reptile, with adult males reaching a maximum of 600cm in length and more than a tonne in weight; females are significantly smaller; head large with relatively narrow but heavy jaws; pair of ridges runs from eyes along centre of snout. *Scales* 2 rows of enlarged nuchal scutes on parietal region; body scales more oval in shape than in other Asian species; anterior nuchals usually absent. *Colouration* Dorsum generally dark greyish, with lighter tans; some individuals have darker bands on flanks and faint bands on body; venter white to creamy-yellow; juveniles yellowish or tan, with black or dark grey spots, and stripes on body and tail. **HABITAT AND HABITS** While no confirmed breeding populations are known, there are occasional records of individuals ranging from small juveniles to mature adults from several coastal parts of Bali. Crocodiles have also been seen and photographed around Nusa Lembongan, Nusa Penida and Nusa Gede. There are records of individuals that escaped or were released from captivity inhabiting few of Bali's internal water bodies. Hatchlings prey on small animals including insects, while adults take larger prey such as deer, buffalo and domestic livestock.

Oviparous, with clutch size of up to 90 eggs laid in mound nests made from vegetation and mud. Female guards nest during incubation period and digs out hatchlings once hatched. **WARNING** This species does carry out unprovoked (and sometimes fatal) attacks on humans, so is considered extremely dangerous. There are historical records of fatal attacks from the island, particularly from the Jembrana regency.

Subadult (top) *and adult female.*

■ CHECKLIST OF THE HERPETOFAUNA OF BALI ■

AMPHIBIANS	
Megophryidae, Litter Frogs	
Hasselt's Litter Frog	*Leptobrachium hasseltii*
Bufonidae, True Toads	
Asian Spined Toad	*Duttaphrynus melanostictus**
Crested Toad	*Ingerophrynus biporcatus*
Microhylidae, Narrow-mouthed Frogs	
Javanese Bullfrog	*Kaloula cf. baleata*
Bali Chorus Frog	*Microhyla orientalis***
Palmated Chorus Frog	*Microhyla palmipes*
Montane Chorus Frog	*Oreophryne monticola*
Dicroglossidae, Fork-tongued Frogs	
Crab-eating Frog	*Fejervarya cancrivora*
Paddy Field Frog	*Fejervarya limnocharis*
Green Puddle Frog	*Occidozyga lima*
Sumatran Puddle Frog	*Occidozyga sumatrana*
Ranidae	
Cricket Frog	*Amnirana nicobariensis*
American Bullfrog	*Aquarana catesbeiana**
White-lipped Frog	*Chalcorana chalconota*
Rhacophoridae, Asian Tree Frogs	
Common Southeast Asian Tree Frog	*Polypedates leucomystax*
TETRAPOD REPTILES	
Emydidae, New World Pond Turtles	
Red-eared Slider	*Trachemys scripta elegans**
Geoemydidae, Asian Hard-shelled Turtles	
Southeast Asian Box Turtle	*Cuora amboinensis*
Asian Leaf Turtle	*Cyclemys dentata*
Black Marsh Turtle	*Siebenrockiella crassicollis**
Trionychidae, Soft-shelled Turtles	
Southeast Asian Soft Terrapin	*Amyda cartilaginea*
Cheloniidae, Hard-shelled Sea Turtles	
Loggerhead Turtle	*Caretta caretta*

Green Turtle	*Chelonia mydas*
Hawksbill Turtle	*Eretmochelys imbricata*
Olive Ridley Turtle	*Lepidochelys olivacea*
Dermochelyidae, Leatherback Sea Turtles	
Leatherback Turtle	*Dermochelys coriacea*
Agamidae, Agamid Lizards/Dragons	
Great Crested Canopy Lizard	*Bronchocela jubata*
Common Garden Lizard	*Calotes versicolor*
Fringed Flying Lizard	*Draco fimbriatus*
Common Flying Lizard	*Draco volans*
Gekkonidae, Cosmopolitan Geckos	
Bent-toed Gecko	*Cyrtodactylus* spp.
Four-clawed Gecko	*Gehyra mutilata*
Tokay Gecko	*Gekko gecko*
Asian House Gecko	*Hemidactylus frenatus*
Flat-tailed Gecko	*Hemidactylus platyurus*
Common Dwarf Gecko	*Hemiphyllodactylus typus*
Lombok Mourning Gecko	*Lepidodactylus lombocensis*
Scincidae, Skinks	
Balinese Snake-eyed Skink	*Cryptoblepharus balinensis*
Beach Snake-eyed Skink	*Cryptoblepharus cursor*
Blue-tailed Snake-eyed Skink	*Cryptoblepharus renschi*
Olive Tree Skink	*Dasia olivacea*
Mangrove Skink	*Emoia atrocostata*
Common Sun Skink	*Eutropis multifasciata*
Rough-scaled Sun Skink	*Eutropis rugifera*
Bowring's Supple Skink	*Lygosoma bowringii*
Short-limbed Supple Skink	*Lygosoma quadrupes*
Yellow-lined Forest Skink	*Sphenomorphus sanctus*
Van Heurn's Forest Skink	*Sphenomorphus vanheurni*
Lacertidae, Wall Lizards	
Asian Grass Lizard	*Takydromus sexlineatus*

Dibamidae, Worm Lizards	
Taylor's Oriental Worm Lizard	*Dibamus taylori*
Varanidae, Monitor Lizards	
Asian Water Monitor	*Varanus salvator*
SNAKES	
Acrochordidae, File Snakes	
Little File Snake	*Acrochordus granulatus*
Pythonidae, Pythons	
Reticulated Python	*Malayopython reticulatus*
Burmese Python	*Python bivittatus*
Xenopeltidae, Sunbeam Snake	
Sunbeam Snake	*Xenopeltis unicolor*
Colubridae, Colubrid Snakes	
Green Vine Snake	*Ahaetulla prasina*
Dog-toothed Cat Snake	*Boiga cynodon*
Mangrove Cat Snake	*Boiga dendrophila**
Marbled Cat Snake	*Boiga multomaculata*
Black-headed Cat Snake	*Boiga nigriceps*
Cuvier's Reed Snake	*Calamaria schlegeli*
Paradise Tree Snake	*Chrysopelea paradisi*
Yellow-striped Racer	*Coelognathus flavolineatus*
Copperhead Racer	*Coelognathus radiatus*
Lesser Sundas Bronzeback	*Dendrelaphis inornatus*
Painted Bronzeback	*Dendrelaphis pictus*
Orange-bellied Snake	*Gongylosoma baliodeirum*
Red-tailed Racer	*Gonyosoma oxycephalum*
Common Wolf Snake	*Lycodon capucinus*
White-banded Wolf Snake	*Lycodon subcinctus*
Boie's Kukri Snake	*Oligodon bitorquatus*
Eight-striped Kukri Snake	*Oligodon octolineatus*
Indo-Chinese Rat Snake	*Ptyas korros*
Banded Rat Snake	*Ptyas mucosa*
Spotted Keelback	*Rhabdophis chrysargos*

Striped Litter Snake	*Sibynophis geminatus*
Javan Keelback	*Xenochrophis melanozostus*
Red-sided Keelback	*Xenochrophis trianguligerus*
Striped Keelback	*Xenochrophis vittatus*
Lamprophiidae, Sand Snakes, House Snakes and similar	
Common Mock Viper	*Psammodynastes pulverulentus*
Indo-Chinese Sand Snake	*Psammophis indochinensis*
Elapidae, Elapids	
Malayan Krait	*Bungarus candidus*
Banded Krait	*Bungarus fasciatus*
Malayan Striped Coral Snake	*Calliophis intestinalis*
Southern Indonesian Spitting Cobra	*Naja sputatrix*
King Cobra	*Ophiophagus Hannah*
Yellow-lipped Sea Krait	*Laticauda colubrina*
Brown-lipped Sea Krait	*Laticauda laticaudata*
Homalopsidae, Mangrove Snakes	
Schneider's Bockadam	*Cerberus schneiderii*
Olive Water Snake	*Hypsiscopus plumbea*
Pareidae, Asian Slug-eaters	
Keeled Slug-eater	*Pareas carinatus*
Viperidae, Vipers and Pit Vipers	
Lesser Sundas White-lipped Viper	*Trimeresurus insularis*
Gerrhopilidae and Typhlopidae, Blind Snakes	
Flowerpot Snake	*Indotyphlops braminus*
Black Blind Snake	*Gerrhopilus ater*
Lesser Sundas Blind Snake	*Sundatyphlops polygrammicus*
CROCODILES	
Crocodylidae, True Crocodiles	
Saltwater Crocodile	*Crocodylus porosus*

*Introduced or possibly introduced species.
**Endemic species.

Further Information

Several books and online resources were extremely useful in the preparation of this book. A few selected references are listed below for further information. Additionally, numerous highly informative, peer-reviewed and general journal articles on the herpetofauna of Indonesia in general and Bali in particular exist. Some of these are available for free through websites such as Google Scholar, JURN and OAJSE.

BOOKS

Whitten, A., Soeriaatmadja, R. and Afiff, S.e.A. (1996). *The Ecology of Java and Bali*. The *Ecology of Indonesia* series Vol II. Periplus Editions, Singapore.

Iskanda, D.T. (1998). *The Amphibians of Java and Bali*. Research and Development Centre for Biology- LIPI, Bogor, Indonesia.

McKay, J.L. (2006). *A field guide to the Amphibians and Reptiles of Bali*. Krieger Publishing Company, Florida, USA.

McKay, J.L. (2006). *Reptil dan Amphibi di Bali*. Krieger Publishing Company, Florida, USA.

Das, I. (2010). *A Field Guide to the Reptiles of South-east Asia*. New Holland Publishers, London, UK.

Das, I. (2012). *A Naturalist's Guide to the Snakes of South-East Asia: Malaysia, Singapore, Thailand, Myanmar, Borneo, Sumatra, Java and Bali*. John Beaufoy Publishing Ltd, Oxford, UK.

Putra, A. and Slade, A. (2015). Snakes: Friends of Farmers in Bali. Bali Reptile Rescue, Bali, Indonesia.

De Lang, R. (2017): *The Snakes of Java, Bali and Surrounding Islands*. Chimaira, Frankfurt, Germany.

WEBSITES

The EMBL/EBI Reptile Database - www.reptile-database.org

AmphibiaWeb - www.amphibiaweb.org

The IUCN Redlist of Threatened Species - www.iucnredlist.org

iNaturalist - www.inaturalist.org

EcologyAsia - www.ecologyasia.com

Appendix

IUCN THREATENED CATEGORIES

Critically Endangered (CR) A taxon is Critically Endangered when it is facing an extremely high risk of extinction in the wild in the near future.

Endangered (EN) A taxon is Endangered when it is facing a very high risk of extinction in the wild in the near future.

Vulnerable (VU) A taxon is Vulnerable when it is facing a high risk of extinction in the wild in the near future.

Near Threatened (NT) A taxon is Near Threatened when it has been evaluated against the criteria but does not qualify for Critically Endangered, Endangered or Vulnerable now, but is close to qualifying for or is likely to qualify for a threatened category in the near future.

Least Concern (LC) A taxon is Least Concern when it has been evaluated against the criteria and does not qualify for Critically Endangered, Endangered, Vulnerable or Near Threatened. Widespread and abundant taxa are included in this category.

Data Deficient (DD) A taxon is Data Deficient when there is inadequate information to make a direct, or indirect, assessment of its risk of extinction based on its distribution and/or population status. A taxon in this category may be well studied, and its biology may be well known, but appropriate data on abundance and/or distribution are lacking.

CITES APPENDICES

Appendix I Contains species threatened with extinction and CITES prohibits international trade in specimens of these species except when the purpose of the import is not commercial, for instance for scientific research. In these exceptional cases, trade may take place provided it is authorized by the granting of both an import permit and an export permit (or re-export certificate).

Appendix II These are species that are not necessarily currently threatened with extinction, but that may become so unless trade is closely controlled. The category also includes so-called 'look-alike species', that is, species of which the specimens in trade look like those of species listed for conservation reasons. International trade in specimens of Appendix II species may be authorized by the granting of an export permit or re-export certificate. No import permit is necessary for these species under CITES (although a permit is needed in some countries that have taken stricter measures than CITES requires). Permits or certificates should only be granted if the relevant authorities are satisfied that certain conditions are met, above all that trade will not detrimental to the survival of the species in the wild.

ACKNOWLEDGEMENTS

This book is a 'group achievement' and many people need to be thanked. First, two of my good old friends, Lindley McKay for making me interested in Balinese herps and Anslem de Silva for the initial suggestion for a book on Bali herps. Then my contemporary fellow herpetologists for their own work, without which mine would have been much more difficult. Then all the beautiful critters featured within. Thanks for tolerating all the flashes and handling, and for posing beautifully for our pictures.

My field work in Bali and Nusa islands would not have been possible or as pleasurable without the help and association of Thumith Nangalla, Samantha Tesoriero, Agus Putra, Puveanthan Govendan, Bayu Wirayudha, Shy Lee, Vion Keraf, Elang Reza, Ajiz Azis, Ketut Arsana, Sumadi Ni Wayan, Yedija Putra Kusuma, Made Bombom Widana and the members of the Friends of the National Parks Foundation (FNPF) of Bali. Lindley, Bayu, Amy Pierce and Ron Lilley helped to build links with the local herpetology community. I am indebted to my wife Nilu for patiently putting up with a busy and unattentive husband at home. Your continuous support has been crucial for the completion of these projects.

I am most thankful to Hastin Ambar Asti, Daicus Belabut, Indraneil Das, Philippe Ganz, Jakob Hallermann, Max Jackson, Thushan Kapurusinghe, Jannico Kelk, Ron Lilley, Ulrich Manthey, Brad Maryan, Lindley McKay, Sven Mecke, Mark O'Shea, Agus Putra, Arne Rasmussen, Nathan Rusli, Kate Sanders, Eric Smith, Wayan Somabawa, Jasmine Vink, Gernot Vogel, Harold Voris, and Made Bombom Widana for the use of their beautiful photographs of living herps and locations. Thasun Amarasinghe, Indraneil Das, Philippe Ganz, Djoko Iskandar, Kalana Maduwage, Lindley McKay, Mizra Kusrini, Jim McGuire, Sven Mecke, Eric Pui Yong Min, Alex Pyron, Arne Rasmussen, Jodi Rowley, Nathan Rusli, Kate Sanders, Glenn Shea, Anjana Silva, Panupong Thammachoti, Gernot Vogel and the CSIRO Library Services helped in sourcing literature and/or providing comments and views. Support for gaining museum records and examining museum specimens came from Paul Doughty, Rebecca Bray and Leanne Griffiths.

Lindley and Indraneil dedicated a lot of time to reviewing the initial manuscript and providing comprehensive feedback, while Krystyna Mayer edited the text. I have addressed most of the remarks, and absolve them of any responsibility for errors that remain. Finally, and perhaps most importantly, I am grateful to John Beaufoy and Rosemary Wilkinson for inviting me to publish this work.

■ INDEX ■

Photo credits: Photos are denoted by a page number followed where relevant by t (top), m (middle), b (bottom), l (left), r (right) or i (inset).

Hastin Ambar Asti 44i; **Daicus Belabut** 110b; **Indraneil Das** 40, 52tr, 66, 113i; **Philippe Ganz** 68; **Jakob Hallermann** 144b; **Max Jackson** 72tr, 97b, 100, 104t, 115t, 117, 119, 124b, 126t, 128tr, 137, 141l, 156b; **Thushan Kapurusinghe** 61, 62; **Jannico Kelk** 128b, 141r; **Ron Lilley** 111, 118, 122, 128tl; **Aaron Lobo** 94, 145br, 149tr & b, 150t; **Ulrich Manthey** 154t; **Brad Maryan** 144tr, 144mr, 148tr; **Lindley McKay** front cover (bm, br), 28, 31bl, 33, 34, 35b, 36, 41t, 46, 76, 79, 80, 81, 82, 85tl, 86, 88, 91, 164br; **Sven Mecke** 47t, 89; **Mark O'Shea** 163; **Agus Putra** 47bl, 60tl, 70br, 84 (juveniles), 102i, 105t, 107, 108b, 109, 114 , 119i, 121, 126b, 127t, 130, 135, 137i, 159t; **Arne Rasmussen** 147tr, 151b; **Nathan Rusli** 31t, 35t, 41b, 47br, 53t, 54, 55, 87b, 112l; 127b, 129; **Kate Sanders** 144tl, 150b; **Eric Smith** 52tl & b, 138, 154b; **Wayan Somabawa** 53t; **Ruchira Somaweera** contents page, 6, 7, 8, 9, 10, 11, 15, 29, 30, 31br, 32, 37, 38, 39, 42, 43, 44, 45, 48, 49, 50, 51, 56, 57, 58, 59, 60tr, 60bl, 62i, 63, 64, 65, 69, 70l & tr, 71, 72tl & b, 73, 74, 75, 77, 78, 81i, 83, 84 (adults), 85tr & b, 87t, 92, 93, 95, 96, 98, 99, 101, 102, 104b, 105b, 106, 112r, 113, 115b, 116, 120, 123, 124t, 125, 131, 132, 136, 140, 143all, 145, 146all, 147b, 148b, 149tl, 152, 153, 155, 156t, 157, 159br, 160, 161, 162, 164t &bl, 165, back cover; **Jasmine Vink** front cover (t, bl), title page, 14, 67, 97t, 103, 108t, 110t, 117i, 134, 139, 158, 159bl; Gernot Vogel 133; **Harold Voris** 147tl, 149m, 151t; **Made Bombom Widana** 13.